Algorithms for Minimization Without Derivatives

RICHARD P. BRENT

Emeritus Professor
Australian National University
Canberra

Dover Publications, Inc.
Mineola, New York

Bibliographical Note

This Dover edition, first published in 2002 and reissued in 2013, is an unabridged republication of the work originally published by Prentice-Hall, Inc., Englewood Cliffs, New Jersey, in 1973.

Library of Congress Cataloging-in-Publication Data

Brent, R. P. (Richard P.)
   Algorithms for minimization without derivatives / Richard P. Brent.
      p. cm.
   Originally published: Englewood Cliffs, N.J. : Prentice-Hall, c1973, in series: Prentice-Hall series in automatic computation.
   Includes bibliographical references and index.
   ISBN-13: 978-0-486-41998-5 (pbk.)
   ISBN-10: 0-486-41998-3 (pbk.)
   1. Maxima and minima. 2. Algorithms. 3. Approximation theory. I. Title.

QA402.5 .B74 2002
511'.66—dc21

2001047459

Manufactured in the United States by Courier Corporation
41998302
www.doverpublications.com

CONTENTS

PREFACE TO THE DOVER EDITION

Since the first edition of *Algorithms for Minimization without Derivatives* was published in 1973, there has been a great deal of research on algorithms for optimization of functions of several variables, the topic of Chapter 7. Also, techniques for computing derivatives analytically have been refined by my former student Andreas Griewank and others, so there is now less reason to consider algorithms which use only function values.

Despite this progress, much of the book is still relevant. The old and deceptively simple problem of approximating a zero of a function of one variable has not gone away. It often occurs as a component of a larger problem, so the efficiency of the algorithm used is important. For example, in their disproof of the Mertens conjecture, Odlyzko and te Riele needed highly accurate values of 2,000 zeros of the Riemann zeta function. Thus, Chapters 3 and 4 are still useful.

Similarly, the problems of approximating local minima of functions of one variable, or global minima of functions of a small number of variables, are still with us, and the algorithms of Chapters 5 and 6 are still relevant.

When Dover offered to publish this book, which had been out of print for many years, I was happy to accept. My first impulse was to start from the beginning and rewrite it, incorporating the most important advances made in the past thirty years, but this impulse was discarded when I realized the scale of the undertaking. With the aim of producing the Dover edition as quickly as possible, a decision was made to reprint the first edition "warts and all," and to maintain a web site at which corrections, updates, programs, and additional references could be found. The web site is at **www.comlab.ox.ac.uk/oucl/work/richard.brent** and readers are invited to visit it.

I would like to thank my mentor and former thesis advisor Gene Golub for putting me in contact with Dover, and John Grafton of Dover for his advice on several matters related to the production of the Dover edition. Finally, I take the opportunity to thank again all those (unfortunately not all still living) who contributed in various ways to the first edition.

R. P. Brent
Oxford, February 2001

PREFACE

The problem of finding numerical approximations to the zeros and extrema of functions, using hand computation, has a long history. Recently considerable progress has been made in the development of algorithms suitable for use on a digital computer. In this book we suggest improvements to some of these algorithms, extend the mathematical theory behind them, and describe some new algorithms for approximating local and global minima. The unifying thread is that all the algorithms considered depend entirely on sequential function evaluations: no evaluations of derivatives are required. Such algorithms are very useful if derivatives are difficult to evaluate, which is often true in practical problems.

An earlier version of this book appeared as Stanford University Report CS-71-198, *Algorithms for finding zeros and extrema of functions without calculating derivatives,* now out of print. This expanded version is published in the hope that it will interest graduate students and research workers in numerical analysis, computer science, and operations research.

I am greatly indebted to Professors G. E. Forsythe and G. H. Golub for their advice and encouragement during my stay at Stanford. Thanks are due to them and to Professors J. G. Herriot, F. W. Dorr, and C. B. Moler, both for their careful reading of various drafts and for many helpful suggestions. Dr. T. J. Rivlin suggested how to find bounds on polynomials (Chapter 6), and Dr. J. H. Wilkinson introduced me to Dekker's algorithm (Chapter 4). Parts of Chapter 4 appeared in Brent (1971d), and are included in this book by kind permission of the Editor of *The Computer Journal.* Thanks go to

Professor F. Dorr and Dr. I. Sobel for their help in testing some of the algorithms; to Michael Malcolm, Michael Saunders, and Alan George for many interesting discussions; and to Phyllis Winkler for her nearly perfect typing. I am also grateful for the influence of my teachers V. Grenness, H. Smith, Dr. D. Faulkner, Dr. E. Strzelecki, Professors G. Preston, J. Miller, Z. Janko, R. Floyd, D. Knuth, G. Polya, and M. Schiffer.

Deepest thanks go to Erin Brent for her help in obtaining some of the numerical results, testing the algorithms, plotting graphs, reading proofs, and in many other ways.

Finally I wish to thank the Commonwealth Scientific and Industrial Research Organization, Australia, for its generous support during my stay at Stanford.

This work is dedicated to Oscar and Nancy Brent, who laid the foundations; and to George Forsythe, who guided the construction.

R. Brent

Algorithms
for Minimization
Without Derivatives

INTRODUCTION
AND
SUMMARY

Consider the problem of finding an approximate zero or minimum of a function of one real variable, using limited-precision arithmetic on a sequential digital computer. The function f may not be differentiable, or the derivative f' may be difficult to compute, so a method which uses only computed values of f is desirable. Since an evaluation of f may be very expensive in terms of computer time, a good method should guarantee to find a correct solution, to within some prescribed tolerance, using only a small number of function evaluations. Hence, we study algorithms which depend on evaluating f at a small number of points, and for which certain desirable properties are guaranteed, even in the presence of rounding errors.

Slow, safe algorithms are seldom preferred in practice to fast algorithms which may occasionally fail. Thus, we want algorithms which are guaranteed to succeed in a reasonable time even for the most "difficult" functions, yet are as fast as commonly used algorithms for "easy" functions. For example, bisection is a safe method for finding a zero of a function which changes sign in a given interval, but from our point of view it is not an acceptable method, because it is just as slow for any function, no matter how well behaved, as it is in the worst possible case (ignoring the possibility that an exact zero may occasionally be found by chance). As a contrasting example, consider the method of successive linear interpolation, which converges superlinearly to a simple zero of a C^1 function, provided that the initial approximation is good and rounding errors are unimportant. This method is not acceptable either, for in practice there may be no way of knowing in advance if the zero is simple, if the initial approximation is sufficiently good to ensure convergence, or if the effect of rounding errors is important.

In Chapter 4 we describe an algorithm which, by combining some of the desirable features of bisection and successive linear interpolation, does come close to satisfying our requirements: it is guaranteed to converge (i.e., halt) after a reasonably small number of function evaluations, and the rate of convergence for well-behaved functions is so fast that a less reliable algorithm is unlikely to be preferred on grounds of speed.

An analogous algorithm, which finds a local minimum of a function of one variable by a combination of golden section search and successive parabolic interpolation, is described in Chapter 5. This algorithm fails to satisfy one of our requirements: in certain applications where repeated one-dimensional minimizations are required, and where accuracy is not very important, a faster (though less reliable) method is preferable. One such application, finding local minima of functions of several variables without calculating derivatives, is discussed in Chapter 7. (Note that wherever we consider minima we could equally well consider maxima.)

Most algorithms for minimizing a nonlinear function of one or more variables find, at best, a local minimum. For a function with several local minima, there is no guarantee that the local minimum found is the global (i.e., true or lowest) minimum. Since it is the global minimum which is of interest in most applications, this is a serious practical disadvantage of most minimization algorithms, and our algorithm given in Chapter 5 is no exception. The usual remedy is to try several different starting points and, perhaps, vary some of the parameters of the minimization procedure, in the hope that the lowest local minimum found is the global minimum. This approach is inefficient, as the same local minimum may be found several times. It is also unreliable, for, no matter how many starting points are tried, it is impossible to be quite sure that the global minimum has been found.

In Chapter 6 we discuss the problem of finding the global minimum to within a prescribed tolerance. It is possible to give an algorithm for solving this problem, provided that a little *a priori* information about the function to be minimized is known. We describe an efficient algorithm, applicable if an upper bound on f'' is known, and we show how this algorithm can be used recursively to find the global minimum of a function of several variables. Unfortunately, because the amount of computation involved increases exponentially with the number of variables, the recursive method is practical only for functions of less than four variables. For functions of more variables, we still have to resort to the unreliable "trial and error" method, unless special information about the function to be minimized is available.

Thus, we are led to consider practical methods for finding local (unconstrained) minima of functions of several variables. As before, we consider methods which depend on evaluating the function at a small number of points. Unfortunately, without imposing very strict conditions on the functions to be minimized, it is not possible to guarantee that an n-dimensional minimization algorithm produces results which are correct to within some prescribed tolerance, or that the effect of rounding errors has been taken into account. We have to be satisfied with algorithms which nearly always give correct results for the functions likely to arise in practical applications.

As suggested by the length of our bibliography, there has recently been considerable interest in the unconstrained minimization problem. Thus, we can hardly expect to find a good method which is completely unrelated to the known ones. In Chapter 7 we take one of the better methods which does not use derivatives, that of Powell (1964), and modify it to try to overcome some of the difficulties observed in the literature. Numerical tests suggest that our proposed method is faster than Powell's original method, and just as reliable. It also compares quite well with a different method proposed by Stewart (1967), at least for functions of less than ten variables. (We have few numerical results for non-quadratic functions of ten or more variables.)

ALGOL implementations of all the above algorithms are given. Most testing was done with ALGOL W (Wirth and Hoare (1966)) on IBM 360/67 and 360/91 computers. As ALGOL W is not widely used, we give ALGOL 60 procedures (Naur (1963)), except for the n-dimensional minimization algorithm. FORTRAN subroutines for the one-dimensional zero-finding and minimization algorithms are given in the Appendix.

To recapitulate, we describe algorithms, and give ALGOL procedures, for solving the following problems efficiently, using only function (not derivative) evaluations:

1. Finding a zero of a function of one variable if an interval in which the function changes sign is given;

2. Finding a local minimum of a function of one variable, defined on a given interval;

3. Finding, to within a prescribed tolerance, the global minimum of a function of one or more variables, given upper bounds on the second derivatives;

4. Finding a local minimum of a function of several variables.

For the first three algorithms, rigorous bounds on the error and the number of function evaluations required are established, taking the effect of rounding errors into account. Some results concerning the order of convergence of the first two algorithms, and preliminary results on interpolation and divided differences, are also of interest.

Section 2
SUMMARY

In this section we summarize the main results of the following chapters. A more detailed discussion is given at the appropriate places in each chapter. This summary is intended to serve as a guide to the reader who is interested in some of our results, but not in others. To assist such a reader, an attempt has been made to keep each chapter as self-contained as possible.

Chapter 2

In Chapter 2 we collect some results on Taylor series, Lagrange interpolation, and divided differences. Most of these results are needed in Chapter 3, and the casual reader might prefer to skip Chapter 2 and refer back to it when necessary. Some of the results are similar to classical ones, but instead of assuming that f has $n + 1$ continuous derivatives, we only assume that $f^{(n)}$ is Lipschitz continuous, and the term $f^{(n+1)}(\xi)$ in the classical results is replaced by a number which is bounded in absolute value by a Lipschitz constant. For example, Lemmas 2.3.1, 2.3.2, 2.4.1, and 2.5.1 are of this nature. Since a Lipschitz continuous function is differentiable almost everywhere, these results are not surprising, although they have not been found in the literature, except where references are given. (Sometimes Lipschitz conditions are imposed on the derivatives of functions of several variables: see, for example, Armijo (1966) and McCormick (1969).) The proofs are mostly similar to those for the classical results.

Theorem 2.6.1 is a slight generalization of some results of Ralston (1963, 1965) on differentiating the error in Lagrange interpolation. It is included both for its independent interest, and because it may be used to prove a slightly weaker form of Lemma 3.6.1 for the important case $q = 2$. (A proof along these lines is sketched in Kowalik and Osborne (1968).)

An interesting result of Chapter 2 is Theorem 2.6.2, which gives an expression for the derivative of the error in Lagrange interpolation at the points of interpolation. It is well known that the conclusion of Theorem 2.6.2 holds if f has $n + 1$ continuous derivatives, but Theorem 2.6.2 shows that it is sufficient for f to have n continuous derivatives.

Theorem 2.5.1, which gives an expansion of divided differences, may be regarded as a generalization of Taylor's theorem. It is used several times in Chapter 3: for example, see Theorem 3.4.1 and Lemma 3.6.1. Theorem

2.5.1 is useful for the analysis of interpolation processes whenever the coefficients of the interpolation polynomials can conveniently be expressed in terms of divided differences.

Chapter 3

In Chapter 3 we prove some theorems which provide a theoretical foundation for the algorithms described in Chapters 4 and 5. In particular, we show when the algorithms will converge superlinearly, and what the order (i.e., rate) of convergence will be. For these results the effect of rounding errors is ignored. The reader who is mainly interested in the practical applications of our results might omit Chapter 3, except for the numerical examples (Section 3.9) and the summary (Section 3.10).

So that results concerning successive linear interpolation for finding zeros (used in Chapter 4), and successive parabolic interpolation for finding turning points (used in Chapter 5), can be given together, we consider a more general process for finding a zero ζ of $f^{(q-1)}$, for any fixed $q \geq 1$. Successive linear interpolation and successive parabolic interpolation are just the special cases $q = 1$ and $q = 2$. Another case which is of some practical interest is $q = 3$, for finding inflexion points. As the proofs for general q are essentially no more difficult than for $q = 2$, most of our results are given for general q.

For the applications in Chapters 4 and 5, the most important results are Theorem 3.4.1, which gives conditions under which convergence is superlinear, and Theorem 3.5.1, which shows when the order is at least $1.618\ldots$ (for $q = 1$) or $1.324\ldots$ (for $q = 2$). These numbers are well known, but our assumptions about the differentiability of f are weaker than those of previous authors, e.g., Ostrowski (1966) and Jarratt (1967, 1968).

From a mathematical point of view, the most interesting result of Chapter 3 is Theorem 3.7.1. The result for $q = 1$ is given in Ostrowski (1966), except for our slightly weaker assumption about the smoothness of f. For $q = 2$, our result that convergence to ζ with order at least $1.378\ldots$ is possible, even if $f^{(3)}(\zeta) \neq 0$, appears to be new. Jarratt (1967) and Kowalik and Osborne (1968) assume that

$$\lim_{n \to \infty} \frac{|x_{n+1} - \zeta|}{|x_n - \zeta|} = 0, \tag{2.1}$$

and then, from Lemma 3.6.1, the order of convergence is $1.324\ldots$. However, even for such a simple function as

$$f(x) = 2x^3 + x^2, \tag{2.2}$$

there are starting points x_0, x_1, and x_2, arbitrarily close to ζ, such that (2.1) fails to hold, and then the order is at least $1.378\ldots$. We should point out that this exceptional case is unlikely to occur: an interesting conjecture is that the set of starting points for which it occurs has measure zero.

The practical conclusion to be drawn from Theorem 3.7.1 is that, if convergence is to be accelerated, then the result of Lemma 3.6.1 should be used in preference to a result like equation (3.2.1). In Section 3.8 we give one of the many ways in which this may be done. Finally, some numerical examples, illustrating both the accelerated and unaccelerated processes, are given in Section 3.9.

Chapter 4

In Chapter 4 we describe an algorithm for finding a zero of a function which changes sign in a given interval. The algorithm is based on a combination of successive linear interpolation and bisection, in much the same way as "Dekker's algorithm" (van Wijngaarden, Zonneveld, and Dijkstra (1963); Wilkinson (1967); Peters and Wilkinson (1969); and Dekker (1969)). Our algorithm never converges much more slowly than bisection, whereas Dekker's algorithm may converge extremely slowly in certain cases. (Examples are given in Section 4.2.)

It is well known that bisection is the optimal algorithm, in a minimax sense, for finding zeros of functions which change sign in an interval. (We only consider sequential algorithms: see Robbins (1952), Wilde (1964), and Section 4.5.) The motivation for both our algorithm and Dekker's is that bisection is not optimal if the class of allowable functions is suitably restricted. For example, it is not optimal for convex functions (Bellman and Dreyfus (1962), Gross and Johnson (1959)), or for C^1 functions with simple zeros.

Both our algorithm and Dekker's exhibit superlinear convergence to a simple zero of a C^1 function, for eventually only linear interpolations are performed and the theorems of Chapter 3 are applicable. Thus, convergence is usually much faster than for bisection. Our algorithm incorporates inverse quadratic interpolation as well as linear interpolation, so it is often slightly faster than Dekker's algorithm on well-behaved functions: see Section 4.4.

Chapter 5

An algorithm for finding a local minimum of a function of one variable is described in Chapter 5. The algorithm combines golden section search (Bellman (1957), Kiefer (1953), Wilde (1964), Witzgall (1969)) and successive parabolic interpolation, in the same way as bisection and successive linear interpolation are combined in the zero-finding algorithm of Chapter 4. Convergence in a reasonable number of function evaluations is guaranteed (Section 5.5). For a C^2 function with positive curvature at the minimum, the results of Chapter 3 show that convergence is superlinear, provided that the minimum is at an interior point of the interval. Other algorithms given in the literature either fail to have these two desirable properties, or their order of convergence is less than that of our algorithm when convergence is strictly superlinear: see Sections 5.4 and 5.5.

In Sections 5.2 and 5.3 we consider the effect of rounding errors. Section 5.2 contains an analysis of the limitations on the accuracy of any algorithm which is based entirely on limited-precision function evaluations, and this section should be studied by the reader who intends to use the ALGOL procedure given in Section 5.8.

If f is unimodal, then our algorithm will find the unique minimum, provided there are no rounding errors. To study the effect of rounding errors, we define "δ-unimodal" functions. A unimodal function is δ-unimodal for all $\delta \geq 0$, but a computed approximation to a unimodal function is not unimodal: it is δ-unimodal for some positive δ, the size of δ depending on the function and on the precision of computation. ($\delta \to 0$ as the precision increases indefinitely.) We prove some theorems about δ-unimodal functions, and give an upper bound on the error in the approximate minimum which is found by our algorithm. In this way we can justify the use of our algorithm in the presence of rounding errors, and account for their effect. Our motivation is rather similar to that of Richman (1968) in developing the ϵ-calculus, but we are not concerned with properties that hold as $\epsilon \to 0$. The reader who is not interested in the effect of rounding errors can skip Section 5.3.

Chapter 6

In Chapter 6 we consider the problem of finding an approximation to the global minimum of a function f, defined on a finite interval, if some *a priori* information about f is given. This interesting problem does not seem to have received much attention, although there have been some empirical investigations (Magee (1960)). In Section 6.1, we show why some *a priori* information is necessary, and discuss some of the possibilities. In the remainder of the chapter we suppose that the information is an upper bound on f''.

An algorithm for global minimization of a function of one variable, applicable when an upper bound on f'' is known, is described in Section 6.3. The basic idea of this algorithm is used by Rivlin (1970) to find bounds on a polynomial in a given interval. We pay particular attention to the problem of giving guaranteed bounds in the presence of rounding errors, and the casual reader may find the details in the last half of Section 6.3 rather indigestible.

In Section 6.4, we try to obtain some insight into the behavior of our algorithm by considering some tractable special cases. Then, in Sections 6.5 and 6.6, we show that no algorithm which uses only function evaluations and an upper bound on f'' could be much faster than our algorithm. Finally, a generalization to functions of several variables is given in Section 6.8. The conditions on f are much weaker than unimodality (Newman (1965)). The generalization is not practically useful for functions of more than three variables, and it is an open question whether a significantly better algorithm for functions of several variables is possible.

Chapter 7

In Chapter 7 we describe a modification of Powell's (1964) algorithm for finding a local minimum of a function of several variables without calculating derivatives. The modification is designed to ensure quadratic convergence, and to avoid the difficulties with Powell's criterion for accepting new search directions.

First, in Section 7.1, we give a brief introduction and a survey of the recent literature. The effect of rounding errors on the attainable accuracy is discussed in Section 7.2. Powell's algorithm is described in Section 7.3, and our main modification is given in Section 7.4. The idea of the modification (finding the principal axes of an approximating quadratic form) is not new: for example, it is used by Greenstadt (1967) in his quasi-Newton method. Unlike Greenstadt, though, we do not use an explicit approximation to the Hessian matrix. An interesting feature of our modification is that it is possible to avoid squaring the condition number of the eigenvalue problem by using a singular value decomposition: see Section 7.4 for the details.

In Sections 7.5 and 7.6 we describe some additional features of our algorithm. Then, in Section 7.7, we give the results of some numerical experiments, and compare our method with those of Powell (1964); Davies, Swann, and Campey (Swann (1964)); and Stewart (1967). For the comparison we have used numerical results obtained by Fletcher (1965) and Stewart (1967). The numerical results suggest that our algorithm is competitive with other current algorithms which do not use derivatives explicitly, although it is difficult to reach a definite conclusion without more practical experience.

Finally, we give a bibliography of some of the recent literature on nonlinear minimization, with an emphasis on methods for solving unconstrained problems. The Appendix contains FORTRAN translations of the ALGOL procedures given in Chapters 4 to 6.

2

SOME USEFUL RESULTS ON TAYLOR SERIES, DIVIDED DIFFERENCES, AND LAGRANGE INTERPOLATION

Section 1
INTRODUCTION

In this chapter we collect some results which are needed in Chapters 3 and 6. The reader who is mainly interested in the practical applications described in Chapters 4 to 7 might prefer to skip this chapter, except for Section 2, and refer back to it when necessary.

Some classical expressions for the error in truncated Taylor series and Lagrange interpolation involve a term $f^{(n+1)}(\xi)$, where ξ is an unknown point. For such expressions to be valid, f must have $n+1$ derivatives. Several of the results of this chapter give expressions which are valid if $f^{(n)}$ satisfies a (possibly one-sided) Lipschitz condition. In these results, the term $f^{(n+1)}(\xi)$ is replaced by a number which is bounded by a Lipschitz constant. It seems unlikely that these results are new, but they have not been found in the literature except where references are given.

The results of Chapter 3 depend heavily on Theorem 5.1, which gives an expansion of the divided difference $f[x_0, \ldots, x_n]$ (Section 2) near the origin. This theorem, and the less cumbersome Corollary 5.1, are useful for the analysis of interpolation processes when the coefficients of the interpolating polynomials can be expressed in terms of divided differences.

Finally, in Section 6, we extend some results of Ralston (1963) on the derivative of the error term in Lagrange interpolation. These results are relevant to Chapter 3, although they are given mainly for their independent interest. Perhaps the most interesting result is Theorem 6.2, which shows that, if we are only concerned with the points of interpolation, then we can differentiate the classical expression for the error (equation (6.4)), regarding the term $f^{(n)}(\xi(x))$ as a constant. This is well known if f has $n + 1$ continuous derivatives, but Theorem 6.2 shows that it is sufficient for f to have n continuous derivatives.

Section 2
NOTATION AND DEFINITIONS

Throughout this chapter $[a, b]$ is a nonempty, finite, closed interval, and f is a real-valued function defined on $[a, b]$. n is a nonnegative integer, M a nonnegative real number, and α a number in $(0, 1]$.

Definitions

The *modulus of continuity* $w(f; \delta)$ of f in $[a, b]$ is defined, for $\delta \geq 0$, by

$$w(f; \delta) = \sup_{\substack{x, y \in [a, b] \\ |x - y| \leq \delta}} |f(x) - f(y)|. \tag{2.1}$$

If f has a continuous n-th derivative on $[a, b]$, then we write $f \in C^n[a, b]$. If, in addition, $f^{(n)} \in Lip_M \alpha$, i.e.,

$$w(f^{(n)}; \delta) \leq M\delta^\alpha \tag{2.2}$$

for all $\delta > 0$, then we write $f \in LC^n[a, b; M, \alpha]$. (This notation is not standard, but it is convenient if we want to mention the constants M and α explicitly.) If $f \in LC^n[a, b; M, 1]$ then we write simply $f \in LC^n[a, b; M]$.

If $x_0, \ldots, x_n$ are distinct points in $[a, b]$, then $IP(f; x_0, \ldots, x_n)$ is the Lagrange interpolating polynomial, i.e., the unique polynomial of degree n or less which coincides with f at $x_0, \ldots, x_n$. The *divided difference* $f[x_0, \ldots, x_n]$ is defined by

$$f[x_0, \ldots, x_n] = \sum_{j=0}^{n} \left(\frac{f(x_j)}{\displaystyle\prod_{\substack{i=0 \\ i \neq j}}^{n} (x_j - x_i)} \right). \tag{2.3}$$

(There are many other notations: see, for example, Milne (1949), Milne-Thomson (1933), and Traub (1964).) Note that, although we suppose for simplicity that $x_0, \ldots, x_n$ are distinct, nearly all the results given here and in Chapter 3 hold if some of $x_0, \ldots, x_n$ coincide. (We then have Hermite interpolation and confluent divided differences: see Traub (1964).) For the statement of these results, the word "distinct" is enclosed in parentheses.

Newton's identities

For future reference, we note the following useful identities (see Cauchy (1840), Isaacson and Keller (1966), or Traub (1964)). The first is often used as the definition of the divided difference $f[x_0, \ldots, x_n]$, while the second gives an explicit representation of the interpolating polynomial and remainder.

1. $f[x_0] = f(x_0)$ and, for $n \geq 1$,

$$f[x_0, \ldots, x_n] = \frac{f[x_0, \ldots, x_{n-1}] - f[x_1, \ldots, x_n]}{x_0 - x_n}. \tag{2.4}$$

2. If $P = IP(f; x_0, \ldots, x_n)$, then

$$f(x) = P(x) + \Big(\prod_{i=0}^{n} (x - x_i)\Big)f[x_0, \ldots, x_n, x], \tag{2.5}$$

and

$$P(x) = f[x_0] + (x - x_0)f[x_0, x_1] + \cdots$$
$$+ (x - x_0) \cdots (x - x_{n-1})f[x_0, \ldots, x_n]. \tag{2.6}$$

Section 3
TRUNCATED TAYLOR SERIES

In this section we give some forms of Taylor's theorem. Lemma 3.1 is needed in Chapter 6, and applies if $f^{(n)}$ satisfies a one-sided Lipschitz condition.

LEMMA 3.1

Suppose that $f \in C^n[0, b]$ for some $b > 0$, and that there is a constant M such that, for all $y \in [0, b]$,

$$f^{(n)}(y) - f^{(n)}(0) \leq My. \tag{3.1}$$

Then, for all $x \in [0, b]$,

$$f(x) = \sum_{r=0}^{n} \frac{x^r}{r!} f^{(r)}(0) + \frac{x^{n+1}}{(n+1)!} m(x), \tag{3.2}$$

where

$$m(x) \leq M. \tag{3.3}$$

Remarks

The proof is by induction on n, and is omitted. The corresponding two-sided result is immediate, and is generalized in Lemma 3.2 below. In Lemma 3.2, fractional factorials are defined in the usual way, so

$$\frac{(n + \alpha)!}{\alpha!} = (1 + \alpha)(2 + \alpha) \cdots (n + \alpha). \tag{3.4}$$

LEMMA 3.2

If $f \in LC^n[a, b; M, \alpha]$ and $x, y \in [a, b]$, then

$$f(x) = \sum_{r=0}^{n} \frac{(x - y)^r}{r!} f^{(r)}(y) + \frac{|x - y|^{n+\alpha} m(x, y)\alpha!}{(n + \alpha)!}, \qquad (3.5)$$

where

$$|m(x, y)| \le M. \qquad (3.6)$$

Remarks

The result is trivial if $n = 0$, and for $n \ge 1$ it follows from Taylor's theorem with the integral form for the remainder, using the integral

$$\int_0^x \frac{t^\alpha (x - t)^{n-1}}{(n - 1)!} dt = \frac{x^{n+\alpha} \alpha!}{(n + \alpha)!} \qquad (3.7)$$

for $x \ge 0$.

Note that the bound (3.6) is sharp, as can be seen from the example

$$f(x) = x^{n+\alpha}, \qquad (3.8)$$

with $y = 0$ and $M = (n + \alpha)!/\alpha!$. Since, for $n \ge 1$,

$$n! < \frac{(n + \alpha)!}{\alpha!}, \qquad (3.9)$$

the bound obtained from the classical result

$$f(x) = \sum_{r=0}^{n-1} \frac{(x - y)^r}{r!} f^{(r)}(y) + \frac{(x - y)^n}{n!} f^{(n)}(\xi), \qquad (3.10)$$

for some ξ between x and y, is not sharp.

Section 4
LAGRANGE INTERPOLATION

The following lemma, used in Chapter 6, gives a one-sided bound on the error in Lagrange interpolation if $f^{(n)}$ satisfies a one-sided Lipschitz condition. Thus, it is similar to Lemma 3.1. The corresponding two-sided result follows from Theorem 3 of Baker (1970), but the proof given here is simpler, and similar to the usual proof of the classical result that, if $f \in C^{n+1}[a, b]$, then $m(x) = f^{(n+1)}(\xi(x))$, for some $\xi(x) \in [a, b]$. (See, for example, Isaacson and Keller (1966), pg. 190.)

LEMMA 4.1

Suppose that $f \in C^n[a, b]$; $x_0, \ldots, x_n$ are (distinct) points in $[a, b]$; $P = IP(f; x_0, \ldots, x_n)$; and, for all $x, y \in [a, b]$ with $x > y$,

$$f^{(n)}(x) - f^{(n)}(y) \le M(x - y). \qquad (4.1)$$

Then, for all $x \in [a, b]$,

$$f(x) = P(x) + \left(\prod_{r=0}^{n} (x - x_r) \right) \frac{m(x)}{(n + 1)!}, \qquad (4.2)$$

where

$$m(x) \leq M. \tag{4.3}$$

Proof

Suppose that $n > 0$ and $x \neq x_r$ for any $r = 0, \ldots, n$, for otherwise the result is trivial. Let

$$w(x) = \prod_{r=0}^{n} (x - x_r), \tag{4.4}$$

and write

$$f(x) = P(x) + w(x)S(x). \tag{4.5}$$

Regarding x as fixed, define

$$F(z) = f(z) - P(z) - w(z)S(x) \tag{4.6}$$

for $z \in [a, b]$.

Thus $F \in C^n[a, b]$, and $F(z)$ vanishes at the $n + 2$ distinct points $x, x_0, \ldots, x_n$. Applying Rolle's theorem n times shows that there are two distinct points $\xi_0, \xi_1 \in (a, b)$ such that

$$F^{(n)}(\xi_0) = F^{(n)}(\xi_1) = 0. \tag{4.7}$$

Differentiating (4.6) n times gives

$$F^{(n)}(z) = f^{(n)}(z) - (n + 1)! \, S(x)z + c(x), \tag{4.8}$$

where $c(x)$ is independent of z. Thus, from (4.7),

$$S(x) = \frac{1}{(n + 1)!} \left[\frac{f^{(n)}(\xi_0) - f^{(n)}(\xi_1)}{\xi_0 - \xi_1} \right], \tag{4.9}$$

and the result follows from condition (4.1).

Section 5
DIVIDED DIFFERENCES

Lemma 5.1 and Theorem 5.1 are needed in Chapter 3. The first part of Lemma 5.1 follows immediately from Lemma 4.1 and the identity (2.5) (we state the two-sided result for variety). The second part is well known, and follows similarly. Theorem 5.1 is more interesting, and most of the results of Chapter 3 depend on it. It may be regarded as a generalization of Taylor's theorem, which is the special case $n = 0$.

LEMMA 5.1

Suppose that $f \in LC^n[a, b; M]$ and that $x_0, \ldots, x_{n+1}$ are (distinct) points in $[a, b]$. Then

$$f[x_0, \ldots, x_{n+1}] = \frac{m}{(n + 1)!}, \tag{5.1}$$

where

$$|m| \leq M. \tag{5.2}$$

Furthermore, if $f \in C^{n+1}[a, b]$, then

$$m = f^{(n+1)}(\xi) \tag{5.3}$$

for some $\xi \in [a, b]$.

THEOREM 5.1

Suppose that $k, n \geq 0$; $f \in C^{n+k}[a, b]$; $a \leq 0$; $b \geq 0$; and $x_0, \ldots, x_n$ are (distinct) points in $[a, b]$. Then

$$f[x_0, \ldots, x_n] = \frac{f^{(n)}(0)}{n!} + \left(\sum_{0 \leq r_1 \leq n} x_{r_1} \right) \frac{f^{(n+1)}(0)}{(n+1)!} + \cdots$$
$$+ \left(\sum_{0 \leq r_1 \leq r_2 \leq \cdots \leq r_k \leq n} x_{r_1} \cdots x_{r_k} \right) \frac{f^{(n+k)}(0)}{(n+k)!} + R, \tag{5.4}$$

where

$$R = \frac{1}{(n+k)!} \left(\sum_{0 \leq r_1 \leq r_2 \leq \cdots \leq r_k \leq n} \{x_{r_1} \cdots x_{r_k}[f^{(n+k)}(\xi_{r_1,\ldots,r_k}) - f^{(n+k)}(0)]\} \right) \tag{5.5}$$

for some $\xi_{r_1,\ldots,r_k}$ in the interval spanned by $x_{r_1}, \ldots, x_{r_k}$ and 0.

COROLLARY 5.1

If, in Theorem 5.1,

$$\delta = \max_{r=0,\ldots,n} |x_r|, \tag{5.6}$$

then

$$|R| \leq \frac{\delta^k}{n! k!} w(f^{(n+k)}; \delta). \tag{5.7}$$

Proof of Theorem 5.1

The result for $k = 0$ is immediate from the second part of Lemma 5.1, so suppose that $k > 0$. Take points $y_0, \ldots, y_n$ which are distinct, and distinct from $x_0, \ldots, x_n$. Then

$$f[x_0, \ldots, x_n] - f[y_0, \ldots, y_n]$$
$$= \sum_{r=0}^{n} \{f[x_0, \ldots, x_r, y_{r+1}, \ldots, y_n] - f[x_0, \ldots, x_{r-1}, y_r, \ldots, y_n]\} \tag{5.8}$$
$$= \sum_{r=0}^{n} (x_r - y_r) f[x_0, \ldots, x_r, y_r, \ldots, y_n], \tag{5.9}$$

by the identity (2.4).

We may suppose, by induction on k, that the theorem holds if k is replaced by $k - 1$ and n by $n + 1$. Use this result to expand each term in (5.9), and consider the limit as $y_0, \ldots, y_n$ tend to 0. By the second part of Lemma 5.1, $f[y_0, \ldots, y_n]$ tends to $f^{(n)}(0)/n!$, so the result follows. (Strictly, to show the existence of the points $\xi_{r_1, \ldots, r_k}$, we must add to the inductive hypothesis the result that $f^{(n+k)}(\xi_{r_1, \ldots, r_k})$ is a continuous function of $x_{r_1}, \ldots, x_{r_k}$.)

Corollary 5.1 is immediate, once we note that there are exactly $(n + k)!/(n!k!)$ terms in the sum (5.5).

Section 6
DIFFERENTIATING THE ERROR

The two theorems in this section are concerned with differentiating the error term for Lagrange interpolation. These theorems are not needed later, but are included for their independent interest, and also because they may be used to give alternative proofs of some of the results of Chapter 3: see Kowalik and Osborne (1968), pp. 18-20.

Theorem 6.1 is given by Ralston (1963, 1965) if $f \in C^{n+1}[a, b]$. We state the result under the slightly weaker assumption that $f \in LC^n[a, b; M]$ for some M: the only difference in the conclusion is that Ralston's term $f^{(n+1)}(\eta(x))$ is replaced by $m(x)$, where $|m(x)| \leq M$. The proof is similar to that given by Ralston (1963), and is also similar to the proof of Lemma 6.2 below, so it is omitted.

Theorem 6.2 gives an expression for the derivative of the error at the points of interpolation. If $f \in LC^n[a, b; M]$ then the result follows immediately from Theorem 6.1, but Theorem 6.2 shows that $f \in C^n[a, b]$ is sufficient.

THEOREM 6.1

Suppose that $n \geq 1$; $f \in LC^n[a, b; M]$; $x_0, \ldots, x_{n-1}$ are (distinct) points in $[a, b]$; $w(x) = (x - x_0) \cdots (x - x_{n-1})$; $P = IP(f; x_0, \ldots, x_{n-1})$; and $f(x) = P(x) + R(x)$. Then there are functions $\xi: [a, b] \to [a, b]$ and $m: [a, b] \to [-M, M]$ such that

1. $f^{(n)}(\xi(x))$ is a continuous function of $x \in [a, b]$ (although $\xi(x)$ is not necessarily continuous);
2. $m(x)$ is continuous on $[a, b]$, except possibly at $x_0, \ldots, x_{n-1}$;
3. for all $x \in [a, b]$,

$$R(x) = \frac{w(x)f^{(n)}(\xi(x))}{n!} \tag{6.1}$$

and

$$R'(x) = \frac{w'(x)f^{(n)}(\xi(x))}{n!} + \frac{w(x)m(x)}{(n + 1)!}; \tag{6.2}$$

and

4. if $x \neq x_r$ for $r = 0, \ldots, n - 1$, then

$$\frac{d}{dx} f^{(n)}(\xi(x)) = \frac{m(x)}{n + 1}. \tag{6.3}$$

THEOREM 6.2

Suppose that $n \geq 1$; $f \in C^n[a, b]$; $x_0, \ldots, x_{n-1}$ are (distinct) points in $[a, b]$; $w(x) = (x - x_0) \cdots (x - x_{n-1})$; $P = IP(f; x_0, \ldots, x_{n-1})$; and $f(x) = P(x) + R(x)$. Then there is a function $\xi: [a, b] \to [a, b]$ such that $f^{(n)}(\xi(x))$ is a continuous function of $x \in [a, b]$; for all $x \in [a, b]$,

$$R(x) = \frac{w(x) f^{(n)}(\xi(x))}{n!}; \tag{6.4}$$

and, for $r = 0, \ldots, n - 1$,

$$R'(x_r) = \frac{w'(x_r) f^{(n)}(\xi(x_r))}{n!}. \tag{6.5}$$

Before proving Theorem 6.2, we need some lemmas. Note the similarity between Lemma 6.2 and Theorem 6.1.

LEMMA 6.1

Suppose that $n \geq 1$; $f \in C^n[a, b]$; $x_0, \ldots, x_n$ are distinct points in $[a, b]$; $P = IP(f; x_0, \ldots, x_n)$;

$$\Delta = \max_{x \in [a,b]} |f^{(n)}(x)|; \tag{6.6}$$

and

$$\delta = \max_{0 \leq i \leq j \leq n} |x_i - x_j|. \tag{6.7}$$

Then, for all $x \in [a, b]$,

$$f(x) = P(x) + \left(\prod_{r=0}^{n} (x - x_r) \right) S(x), \tag{6.8}$$

where

$$|S(x)| \leq \frac{2\Delta}{n! \, \delta}. \tag{6.9}$$

Proof

If $x = x_r$ for some $r = 0, \ldots, n$, then we can take $S(x) = 0$. Otherwise, by the identity (2.5),

$$S(x) = f[x_0, \ldots, x_n, x]. \tag{6.10}$$

Write x_{n+1} for x, and reorder $x_0, \ldots, x_{n+1}$ so that, if the reordered points are $x'_0, \ldots, x'_{n+1}$, then

$$x'_0 - x'_{n+1} = \max_{0 \leq i < j \leq n+1} |x'_i - x'_j| \geq \delta. \tag{6.11}$$

From (6.10) and the identity (2.4),

$$S(x) = \frac{f[x_0', \ldots, x_n'] - f[x_1', \ldots, x_{n+1}']}{x_0' - x_{n+1}'}, \tag{6.12}$$

so, by Lemma 5.1,

$$S(x) = \frac{f^{(n)}(\xi) - f^{(n)}(\xi')}{n!(x_0' - x_{n+1}')} \tag{6.13}$$

for some ξ and ξ' in $[a, b]$. In view of (6.6) and (6.11), the result follows.

LEMMA 6.2

Suppose that $n \geq 2$; $f \in C^n[a, b]$; $x_0, \ldots, x_{n-1}$ are distinct points in $[a, b]$; $\Delta = \max\limits_{x \in [a,b]} |f^{(n)}(x)|$; $\delta = \max\limits_{0 \leq i < j < n} |x_i - x_j|$; $P_n = IP(f; x_0, \ldots, x_{n-1})$; $w_n(x) = (x - x_0) \cdots (x - x_{n-1})$; and $f(x) = P_n(x) + R(x)$. Then there is a function $\xi: [a, b] \to [a, b]$ such that, for all $x \in [a, b]$, $f^{(n)}(\xi(x))$ is a continuous function of x;

$$R(x) = \frac{w_n(x)f^{(n)}(\xi(x))}{n!}; \tag{6.14}$$

$$\left| R'(x) - \frac{w_n'(x)f^{(n)}(\xi(x))}{n!} \right| \leq \frac{2|w_n(x)|\Delta}{n!\,\delta}; \tag{6.15}$$

and, if $x \neq x_r$ for $r = 0, \ldots, n - 1$, then

$$\left| \frac{d}{dx} f^{(n)}(\xi(x)) \right| \leq \frac{2\Delta}{\delta}. \tag{6.16}$$

Proof

Let x_n be a point in $[a, b]$, distinct from x and $x_0, \ldots, x_{n-1}$. For $k = n$ or $n + 1$, define

$$P_k = IP(f; x_0, \ldots, x_{k-1}) \tag{6.17}$$

and

$$w_k(x) = (x - x_0) \cdots (x - x_{k-1}). \tag{6.18}$$

By the classical result corresponding to Lemma 4.1, there is a function ξ such that (6.14) holds. Suppose, until further notice, that $x \neq x_r$ for $r = 0, \ldots, n$. Then, from (6.14) and the identity

$$P_k(x) = \sum_{r=0}^{k-1} \frac{f(x_r)w_k(x)}{(x - x_r)w_k'(x_r)}, \tag{6.19}$$

we have

$$\frac{f^{(n)}(\xi(x))}{n!} = \frac{f(x)}{w_n(x)} - \sum_{r=0}^{n-1} \frac{f(x_r)}{(x - x_r)w_n'(x_r)}. \tag{6.20}$$

Since the right side of (6.20) is continuously differentiable at x, so is the left side, and

$$\frac{1}{n!}\frac{d}{dx} f^{(n)}(\xi(x)) = \frac{d}{dx}\left(\frac{f(x)}{w_n(x)}\right) + \sum_{r=0}^{n-1} \frac{f(x_r)}{(x - x_r)^2 w_n'(x_r)}. \tag{6.21}$$

Define $S(x, x_n)$ by

$$f(x) = P_{n+1}(x) + w_{n+1}(x)S(x, x_n). \tag{6.22}$$

Since

$$w'_{n+1}(x_r) = \begin{cases} w_n(x_n) & \text{if } r = n, \\ (x_r - x_n)w'_n(x_r) & \text{if } r = 0, \ldots, n-1, \end{cases} \tag{6.23}$$

equation (6.19) gives

$$\frac{P_{n+1}(x)}{w_{n+1}(x)} = \sum_{r=0}^{n-1} \frac{f(x_r)}{(x - x_r)(x_r - x_n)w'_n(x_r)} + \frac{f(x_n)}{(x - x_n)w_n(x_n)}, \tag{6.24}$$

so

$$S(x, x_n) = \frac{f(x)/w_n(x) - f(x_n)/w_n(x_n)}{x - x_n} + \sum_{r=0}^{n-1} \frac{f(x_r)}{(x - x_r)(x_n - x_r)w'_n(x_r)}. \tag{6.25}$$

As $x_n \to x$, the right side of (6.25) tends to the right side of (6.21). Thus, there exists

$$\lim_{x_n \to x} S(x, x_n) = \frac{1}{n!}\frac{d}{dx}f^{(n)}(\xi(x)), \tag{6.26}$$

and, from the definition (6.22) and Lemma 6.1, this proves (6.16). Now, by differentiating the right side of (6.14) by parts, we see that (6.15) holds; in fact

$$R'(x) = \frac{w'_n(x)f^{(n)}(\xi(x)) + w_n(x)(df^{(n)}(\xi(x))/dx)}{n!}, \tag{6.27}$$

provided that $x \neq x_r$ for $r = 0, \ldots, n-1$. Consider (6.27) near one of the points x_r, $r = 0, \ldots, n-1$. $R'(x)$ is continuous at x_r, $w_n(x_r) = 0$, $w'_n(x_r) \neq 0$, and, by (6.16), $df^{(n)}(\xi(x))/dx$ is bounded for $x \neq x_r$. Thus $f^{(n)}(\xi(x))$ has, at worst, a removable discontinuity at x_r. By the continuity of $f^{(n)}(\xi)$ as a function of ξ, a suitable redefinition of $\xi(x_r)$ will ensure that $f^{(n)}(\xi(x))$ is a continuous function of x, and that

$$R'(x_r) = \frac{w'_n(x_r)f^{(n)}(\xi(x_r))}{n!}. \tag{6.28}$$

This completes the proof of the lemma.

Proof of Theorem 6.2

If $n \geq 2$ then the result follows immediately from Lemma 6.2. If $n = 1$, choose $\xi(x)$ so that $\xi(x_0) = x_0$ and, for $x \neq x_0$,

$$f'(\xi(x)) = \frac{f(x) - f(x_0)}{x - x_0}. \tag{6.29}$$

Then $f'(\xi(x))$ is a continuous function of $x \in [a, b]$ and, as $R(x) = f(x) - f(x_0)$ and $w(x) = x - x_0$, it is easy to see that equations (6.4) and (6.5) are satisfied. Thus, the theorem holds for all $n \geq 1$.

3

THE USE OF SUCCESSIVE INTERPOLATION FOR FINDING SIMPLE ZEROS OF A FUNCTION AND ITS DERIVATIVES

Section 1
INTRODUCTION

Suppose that $q \geq 1$ and $f \in C^{q-1}[a, b]$. Given (distinct) points $x_0, \ldots, x_q$ in $[a, b]$, a sequence (x_n) may be defined in the following way: if $x_0, \ldots, x_{n+q}$ are already defined, let $P_n = IP(f; x_n, \ldots, x_{n+q})$ be the q-th degree polynomial which coincides with f at $x_n, \ldots, x_{n+q}$, and choose x_{n+q+1} so that

$$P_n^{(q-1)}(x_{n+q+1}) = 0. \qquad (1.1)$$

Under certain conditions the sequence (x_n) is well defined by (1.1), lies in $[a, b]$, and converges to a zero ζ of $f^{(q-1)}$. In this chapter we give sufficient conditions for convergence, and estimate the asymptotic rate of convergence, making various assumptions about the differentiability of f.

Since P_n is a polynomial of degree q, (1.1) is a linear equation in x_{n+q+1}. If

$$f[x_n, \ldots, x_{n+q}] \neq 0, \qquad (1.2)$$

then Lemma 3.1 shows that the unique solution is

$$x_{n+q+1} = \frac{1}{q}\left(\sum_{l=1}^{q} x_{n+l} - \frac{f[x_{n+1}, \ldots, x_{n+q}]}{f[x_n, \ldots, x_{n+q}]}\right), \qquad (1.3)$$

and this might be used as an alternative definition. From Section 4 on, our assumptions ensure that $x_n, \ldots, x_{n+q}$ are sufficiently close to a simple zero ζ of $f^{(q-1)}$, so (1.2) holds. In Section 3 the assumption that $f^{(q)}(\zeta) \neq 0$ is unnecessary: all that is required is that x_{n+q+1} is a (not necessarily unique) solution of (1.1).

The cases of most practical interest are $q = 1, 2$, and 3. For $q = 1$, the successive interpolation process reduces to the familiar method of successive linear interpolation for finding a zero of f, and some of our results are well known. (See Collatz (1964), Householder (1970), Ortega and Rheinboldt (1970), Ostrowski (1966), Schröder (1870), and Traub (1964, 1967).) For $q = 2$, we have a process of successive parabolic interpolation for finding a turning point; for $q = 3$, a process for finding an inflexion point. These two cases are discussed separately by Jarratt (1967, 1968), who assumes that f is analytic near ζ. By using (1.3) and Theorem 2.5.1, we show that much milder assumptions on the smoothness of f suffice (Theorems 4.1, 5.1, and 7.1). Also, most of our results hold for any $q \geq 1$, and the proofs are no more difficult than those for the special cases $q = 2$ and $q = 3$.

Some simplifying assumptions

Practical algorithms for finding zeros and extrema, using the results of this chapter, are discussed in Chapters 4 and 5. Until then we ignore the problem of rounding errors, and usually suppose that the initial approximations $x_0, \ldots, x_q$ are sufficiently good.

For the sake of simplicity, we assume that any $q + 1$ consecutive points $x_n, \ldots, x_{n+q}$ are distinct. This is always true in the applications described in Chapters 4 and 5. Thus, P_n is just the Lagrange interpolating polynomial, and the results of Chapter 2 are applicable. As in Chapter 2, the assumption of distinct points is not necessary, and the same results hold without this assumption if P_n is the appropriate Hermite interpolating polynomial.

A preview of the results

The definition of "order of convergence" is discussed in Section 2, and in Section 3 we show that, if a sequence (x_n) satisfies (1.1) and converges to ζ, then $f^{(q-1)}(\zeta) = 0$ (Theorem 3.1).

In Sections 4 to 7, we consider the rate of convergence to a simple zero ζ of $f^{(q-1)}$, making increasingly stronger assumptions about the smoothness of f. For practical applications, the most important result is probably Theorem 4.1, which shows that convergence is superlinear if $f \in C^q$ and the starting values are sufficiently good. As in similar results for Newton's method (Collatz (1964), Kantorovich and Akilov (1959), Ortega (1968), Ortega and Rheinboldt (1970), etc.), it is possible to say precisely what "sufficiently good" means. Theorem 5.1 is an easy consequence of Theorem 4.1, and gives a lower bound on the order of convergence if $f^{(q)}$ is Lipschitz continuous.

The question of when the order of convergence is equal to the lower bound given by Theorem 5.1 is the subject of Sections 6 and 7. Although the results are interesting, they are not of much practical importance, for in practical problems it is merely a pleasant surprise if the iterative process converges faster than expected! Thus, the reader interested mainly in practical applications may prefer to skip Sections 6 and 7 (and also Theorem 3.1), except for Lemma 6.1.

In Section 8, we consider the interesting problem of accelerating the rate of convergence. Theorem 8.1 shows how this may be done. We make use of Lemma 6.1, which gives a recurrence relation for the error in successive approximations to ζ, and is a generalization of results of Ostrowski (1966) and Jarratt (1967, 1968).

Finally, in Section 9 the theoretical results are illustrated by some numerical examples, and a brief summary of the main theorems is given in Section 10. The reader may find it worthwhile to glance at this summary occasionally in order to see the pattern of the results.

Section 2
THE DEFINITION OF ORDER

Suppose that $\lim_{n \to \infty} x_n = \zeta$. There are many reasonable definitions of the "order of convergence" of the sequence (x_n). For example, we could say that the order of convergence is ρ if one or more of (2.1) to (2.4) holds:

$$\lim_{n \to \infty} \frac{|x_{n+1} - \zeta|}{|x_n - \zeta|^\rho} = K > 0, \qquad (2.1)$$

$$\lim_{n \to \infty} \frac{\log |x_{n+1} - \zeta|}{\log |x_n - \zeta|} = \rho, \qquad (2.2)$$

$$\lim_{n \to \infty} (-\log |x_n - \zeta|)^{1/n} = \rho, \qquad (2.3)$$

$$\liminf_{n \to \infty} (-\log |x_n - \zeta|)^{1/n} = \rho. \qquad (2.4)$$

These conditions are in decreasing order of strength, i.e., $(2.1) \supset (2.2) \supset (2.3) \supset (2.4)$, and none of them are equivalent. (2.1) is used by Ostrowski (1966), Jarratt (1967), and Traub (1964, 1967), while (2.2) is used by Wall (1956), Tornheim (1964), and Jarratt (1968). Voigt (1971) and Ortega and Rheinboldt (1970) give some more possibilities. For example, we may take the supremum of ρ such that the limit K in (2.1) exists and is zero, or the infimum of ρ such that K is infinite. For our purposes it is convenient to use (2.1) and (2.4), so we make the following definitions.

DEFINITION 2.1

We say $x_n \to \zeta$ with *strong order* ρ and *asymptotic constant* K if $x_n \to \zeta$ as $n \to \infty$ and (2.1) holds.

We say $x_n \to \zeta$ with *weak order* ρ if $x_n \to \zeta$ as $n \to \infty$ and (2.4) holds. If $x_n = \zeta$ for all sufficiently large n then we say that $x_n \to \zeta$ with weak order ∞.

DEFINITION 2.2

Let

$$c = \limsup_{n \to \infty} |x_n - \zeta|^{1/n}. \tag{2.5}$$

We say $x_n \to \zeta$ *sublinearly* if $x_n \to \zeta$ and $c = 1$. We say $x_n \to \zeta$ *linearly* if $0 < c < 1$. We say $x_n \to \zeta$ *superlinearly* if $c = 0$. We say $x_n \to \zeta$ *strictly superlinearly* if $x_n \to \zeta$ with weak order $\rho > 1$.

Examples

Some remarks and examples may help to clarify the definitions. If $\rho > 1$ and $x_n = \exp(-\rho^n)[1 + o(1)]$ as $n \to \infty$, then $x_n \to 0$ with strong order ρ and asymptotic constant 1. If $\sigma > 1$ and $x_n = \exp(-\sigma^n)[2 + (-1)^n]$, then $x_n \to 0$ with weak order σ, but not with any strong order, for the limit in (2.1) does not exist if $\rho = \sigma$, is zero if $\rho < \sigma$, and is infinite if $\rho > \sigma$. Thus, convergence with strong order ρ implies convergence with weak order ρ, but not conversely.

If the limit in (2.1) or (2.4) exists, and $x_n \to \zeta$, then $\rho \geq 1$. If the limit (2.1) exists with $\rho = 1$, and $x_n \to \zeta$, then $K \leq 1$. ($K < 1$ for linear convergence, and $K = 1$ for sublinear convergence.)

Examples of sublinear, linear, superlinear, and strictly superlinear convergence are $x_n = 1/n$, 2^{-n}, n^{-n}, and 2^{-2^n} respectively.

Section 3

CONVERGENCE TO A ZERO

In this section we show that if the sequence (x_n) defined by (1.1) converges, then it must converge to a zero of $f^{(q-1)}$, assuming only that $f \in C^{q-1}[a, b]$. First, we need a lemma which gives a relation between the points $x_n, \ldots, x_{n+q+1}$.

LEMMA 3.1

If $x_n, x_{n+1}, \ldots, x_{n+q}$ are (distinct) points in $[a, b]$, and x_{n+q+1} satisfies (1.1), then

$$\left(\sum_{i=0}^{q-1} (x_{n+i} - x_{n+q+1}) \right) f[x_n, \ldots, x_{n+q}] = f[x_n, \ldots, x_{n+q-1}]. \tag{3.1}$$

Proof

By the identity (2.2.6),

$$P_n(x) = f[x_n] + (x - x_n)f[x_n, x_{n+1}] + \cdots$$
$$+ (x - x_n) \cdots (x - x_{n+q-1})f[x_n, \ldots, x_{n+q}], \tag{3.2}$$

so

$$P_n^{(q-1)}(x) = (q-1)!\left\{ f[x_n, \ldots, x_{n+q-1}] - \left(\sum_{i=0}^{q-1} (x_{n+i} - x)\right) f[x_n, \ldots, x_{n+q}] \right\}.$$
(3.3)

Thus, the result follows from (1.1).

THEOREM 3.1

Suppose that $f \in C^{q-1}[a, b]$; that a sequence (x_n) satisfying (1.1) is defined in $[a, b]$; and that there exists $\lim_{n\to\infty} x_n = \zeta$. Then $f^{(q-1)}(\zeta) = 0$.

Proof

Suppose, by way of contradiction, that

$$f^{(q-1)}(\zeta) \neq 0.$$
(3.4)

For $0 \leq r < q$, the identity (2.2.4) shows that

$$(x_{n+r} - x_{n+q})f[x_n, \ldots, x_{n+q}]$$
$$= f[x_n, \ldots, x_{n+q-1}] - f[x_n, \ldots, x_{n+r-1}, x_{n+r+1}, \ldots, x_{n+q}].$$
(3.5)

Thus, from Lemma 3.1,

$$x_{n+r} - x_{n+q} = \mu_{n,r} \sum_{i=0}^{q-1} (x_{n+i} - x_{n+q+1}),$$
(3.6)

where

$$\mu_{n,r} = 1 - \frac{f[x_n, \ldots, x_{n+r-1}, x_{n+r+1}, \ldots, x_{n+q}]}{f[x_n, \ldots, x_{n+q-1}]}.$$
(3.7)

Both divided differences in (3.7) tend to $f^{(q-1)}(\zeta)/(q-1)!$ as $n \to \infty$, so there is no loss of generality in assuming that the denominator $f[x_n, \ldots, x_{n+q-1}]$ is nonzero for all n (on the assumption (3.4)), and we have

$$\lim_{n\to\infty} \mu_{n,r} = 0.$$
(3.8)

Summing (3.6) over $r = 0, \ldots, q-1$ and rearranging terms gives

$$\sum_{r=0}^{q-1} (x_{n+r} - x_{n+q+1}) = \mu_n'(x_{n+q} - x_{n+q+1}),$$
(3.9)

where

$$\mu_n' = \frac{q}{1 - \sum_{r=0}^{q-1} \mu_{n,r}}.$$
(3.10)

By (3.8), there is no loss of generality in assuming that the denominator in (3.10) is nonzero for all $n \geq 0$. From (3.6), with $r = q-1$, and (3.9), we have

$$x_{n+q-1} - x_{n+q} = \mu_n(x_{n+q} - x_{n+q+1}),$$
(3.11)

where

$$\mu_n = \mu_{n,q-1}\mu'_n. \qquad (3.12)$$

The repeated application of (3.11) gives

$$x_{q-1} - x_q = \mu_0\mu_1 \cdots \mu_n(x_{n+q} - x_{n+q+1}) \qquad (3.13)$$

and, by (3.8), (3.10) and (3.12), $\mu_n \to 0$ as $n \to \infty$, so the right side of (3.13) tends to zero as $n \to \infty$. This contradicts the assumption that $x_{q-1} \neq x_q$, so (3.4) is false, and the proof is complete. (If we do not wish to assume that any $q + 1$ consecutive points $x_n, \ldots, x_{n+q}$ are distinct, then we may argue as follows: on the assumption (3.4), the right side of (3.1) is nonzero for all sufficiently large n, and thus at least two consecutive points from $x_n, \ldots, x_{n+q+1}$ are distinct. Taking these two points in place of x_{q-1} and x_q, we get a contradiction in the same way as from (3.13).)

Section 4
SUPERLINEAR CONVERGENCE

If f has one more continuous derivative than required in Theorem 3.1, then Theorem 4.1 shows that convergence to a simple zero of $f^{(q-1)}$ is superlinear, in the sense of Definition 2.2, provided the starting values are sufficiently good. The theorem makes precise what we mean by "sufficiently good." (In equation (4.1), w is the modulus of continuity: see Section 2.2.) Convergence to a multiple zero of $f^{(q-1)}$ is not usually superlinear, even if $q = 1$ (Section 4.2), and Theorem 3.1 above is the only theorem in this chapter for which we do not need to assume that the zero is simple. Thus, there is no reason to expect that the algorithms described in Chapters 4 and 5 will converge any faster than linearly to multiple zeros of $f^{(q-1)}$.

THEOREM 4.1
Suppose that $f \in C^q[a, b]$; $\zeta \in [a, b]$; $x_0, \ldots, x_q$ are (distinct) points in $[a, b]$; $\delta_0 = \max\limits_{i=0,\ldots,q} |x_i - \zeta|$; $f^{(q-1)}(\zeta) = 0$; $[\zeta - \delta_0, \zeta + \delta_0] \subseteq [a, b]$; and

$$3w(f^{(q)}; \delta_0) < |f^{(q)}(\zeta)|. \qquad (4.1)$$

Then a sequence (x_n) is uniquely defined by (1.1), and $x_n \to \zeta$ superlinearly as $n \to \infty$. Furthermore, if, for $n \geq 0$,

$$\delta_n = \max\limits_{i=0,\ldots,q} |x_{n+i} - \zeta| \qquad (4.2)$$

and

$$\lambda_n = \frac{3w(f^{(q)}; \delta_n)}{|f^{(q)}(\zeta)|}, \qquad (4.3)$$

then the sequence (δ_n) is monotonic decreasing and

$$\delta_{n+q+1} \leq \lambda_n \delta_{n+1}. \qquad (4.4)$$

Proof

Without loss of generality, assume that $\zeta = 0$. Let δ_n and λ_n be as in the statement of the theorem (equations (4.2) and (4.3)). Since $f^{(q-1)}(0) = 0$, Corollary 2.5.1 to Theorem 2.5.1 (with $k = 1$, $n = q - 1$) gives

$$f[x_1, \ldots, x_q] = \left(\sum_{i=1}^{q} x_i\right) \frac{f^{(q)}(0)}{q!} + R_1, \tag{4.5}$$

where

$$|R_1| \leq \frac{\delta' w(f^{(q)}; \delta')}{(q-1)!} \tag{4.6}$$

if

$$\delta' = \max_{i=1,\ldots,q} |x_i| \leq \delta_0. \tag{4.7}$$

Similarly,

$$f[x_0, \ldots, x_q] = \frac{f^{(q)}(0)}{q!}(1 + R_2) = \frac{f^{(q)}(0)}{q!(1 + R_3)}, \tag{4.8}$$

where

$$|R_2| \leq \frac{w(f^{(q)}; \delta_0)}{|f^{(q)}(0)|} = \frac{\lambda_0}{3} < \frac{1}{3}, \tag{4.9}$$

so

$$|R_3| = \left|\frac{R_2}{1 + R_2}\right| \leq \frac{\lambda_0}{2} < \frac{1}{2}. \tag{4.10}$$

(Note that the assumption (4.1) ensures that $f[x_0, \ldots, x_q] \neq 0$.)

From Lemma 3.1 (with x_0 and x_q interchanged), (4.5), and (4.8),

$$\left(\sum_{i=1}^{q} (x_i - x_{q+1})\right) \frac{f^{(q)}(0)}{q!} = \left(\sum_{i=1}^{q} x_i\right) \frac{f^{(q)}(0)}{q!} + R_4, \tag{4.11}$$

where

$$R_4 = R_3 \left(\sum_{i=1}^{q} x_i\right) \frac{f^{(q)}(0)}{q!} + R_1(1 + R_3). \tag{4.12}$$

From (4.6), (4.7), and (4.10), equation (4.12) gives

$$|R_4| \leq \frac{\lambda_0 \delta' |f^{(q)}(0)|}{2 \cdot (q-1)!} + \frac{3\delta' w(f^{(q)}; \delta')}{2 \cdot (q-1)!}, \tag{4.13}$$

so, from (4.3) and (4.7),

$$|R_4| \leq \frac{\lambda_0 \delta' |f^{(q)}(0)|}{(q-1)!}. \tag{4.14}$$

Now, from (4.11), we have

$$|x_{q+1}| \leq \lambda_0 \delta'. \tag{4.15}$$

By the assumption (4.1), $\lambda_0 < 1$, so x_{q+1} lies in $[a, b]$, δ_1 and λ_1 are well-defined, $\delta_1 = \delta' \leq \delta_0$, $\lambda_1 \leq \lambda_0$, and

$$|x_{q+1}| \leq \lambda_0 \delta_1. \tag{4.16}$$

In the same way, we see that $\delta_0 \geq \delta_1 \geq \delta_2 \geq \cdots, 1 > \lambda_0 \geq \lambda_1 \geq \lambda_2 \geq \cdots$, and, for $n \geq 0$,

$$|x_{n+q+1}| \leq \lambda_n \delta_{n+1}. \tag{4.17}$$

Thus, the inequality (4.4) holds, and it only remains to show that $x_n \to 0$ superlinearly. From (4.4) and the above,

$$\delta_{kq+1} \leq \lambda_0 \lambda_q \cdots \lambda_{(k-1)q} \delta_1 \leq \lambda_0^k \delta_1, \tag{4.18}$$

and $\lambda_0 < 1$ by assumption (4.1), so $\delta_n \to 0$ as $n \to \infty$. Thus, by the continuity of $f^{(q)}$ and the definition (4.3), $\lambda_n \to 0$ as $n \to \infty$.

Take any $\epsilon > 0$. For all sufficiently large n,

$$\lambda_n \leq \epsilon^q, \tag{4.19}$$

so, from (4.4),

$$\limsup_{n \to \infty} \delta_n^{1/n} \leq \epsilon. \tag{4.20}$$

As ϵ is arbitrarily small, this shows that

$$\lim_{n \to \infty} |x_n|^{1/n} = \lim_{n \to \infty} \delta_n^{1/n} = 0. \tag{4.21}$$

Thus, $x_n \to \zeta = 0$ superlinearly, and the proof is complete.

Remarks

The proof of Theorem 4.1 shows that, for $n \geq 0$, $|x_{n+q+1} - \zeta|$ is no greater than the second-largest of $|x_n - \zeta|, \ldots, |x_{n+q} - \zeta|$. Thus, if $q = 1$, the sequence $(|x_n - \zeta|)$ is monotonic decreasing, except perhaps for the first term. In fact, the proof shows that, for $q = 1$ and $n \geq 1$,

$$\frac{|x_{n+1} - \zeta|}{|x_n - \zeta|} \leq \lambda_{n-1} \longrightarrow 0 \text{ as } n \longrightarrow \infty \tag{4.22}$$

(provided $x_n \neq \zeta$). This is a common definition of "superlinear convergence," stronger than our Definition 2.2.

If $q \geq 2$, the sequence $(|x_n - \zeta|)$ need not be eventually monotonic decreasing. Monotonicity would follow from strong superlinear convergence with order greater than 1, but more conditions are necessary to ensure this sort of convergence: see Sections 6 and 7.

Section 5
STRICT SUPERLINEAR CONVERGENCE

Assuming slightly more than Theorem 4.1, Theorem 5.1 shows that convergence to a simple zero of $f^{(q-1)}$ is strictly superlinear (Definition 2.2). Before stating the theorem, we define some constants $\beta_{q,\alpha}$ and $\gamma_{q,\alpha}$ which are needed here and in Sections 6 and 7.

DEFINITION 5.1

For $q \geq 1$ and $\alpha > 0$, let the roots of

$$x^{q+1} = x + \alpha \tag{5.1}$$

be $u_{q,\alpha}^{(i)}$ for $i = 0, \ldots, q$, with $|u_{q,\alpha}^{(0)}| \geq |u_{q,\alpha}^{(1)}| \geq \cdots \geq |u_{q,\alpha}^{(q)}|$. Then the constants $\beta_{q,\alpha}$ and $\gamma_{q,\alpha}$ are defined by

$$\beta_{q,\alpha} = |u_{q,\alpha}^{(0)}| \quad \text{and} \quad \gamma_{q,\alpha} = |u_{q,\alpha}^{(1)}|.$$

Since the case $\alpha = 1$ often occurs, we write simply β_q for $\beta_{q,1}$, and γ_q for $\gamma_{q,1}$.

Remarks

$\beta_{q,\alpha}$ is the unique positive real root of (5.1), and it is easy to see that, for $0 < \alpha \leq 1$,

$$(1 + \alpha)^{2/(2q+1)} < \beta_{q,\alpha} < (1 + \alpha)^{1/q}. \tag{5.2}$$

We are only interested in the constants $\gamma_{q,\alpha}$ when $\alpha = 1$. If $\alpha = 1$ and $q \geq 2$ then there are exactly two complex conjugate roots of (5.1) with modulus γ_q. If $q = 1$ or 2 then $\gamma_q < 1$, but, for $q \geq 3$, $1 < \gamma_q < \beta_q$. This may be proved by applying the Lehmer-Schur test to show that, for suitable $\epsilon > 0$, exactly $q - 2$ roots of

$$x^{q+1} = x + 1 \tag{5.3}$$

lie in the circle $|x| < 1 + \epsilon$. The details are omitted, for all cases of practical interest are covered by Table 5.1, which gives β_q and γ_q to 12 decimal places for $q = 1, \ldots, 10$. The table was computed by finding all roots of (5.3) with the program of Jenkins (1969), and the entries are the correctly rounded values of β_q and γ_q if Jenkins's *a posteriori* error bounds are correct.

TABLE 5.1   The constants $\boldsymbol{\beta}_q$ and $\boldsymbol{\gamma}_q$ for $q = 1(1)10*$

q	β_q	γ_q
1	1.618033988750	0.618033988750
2	1.324717957245	0.868836961833
3	1.220744084606	1.063336938821
4	1.167303978261	1.099000315146
5	1.134724138402	1.099174913506
6	1.112775684279	1.091953305766
7	1.096981557799	1.083743696285
8	1.085070245491	1.076133134033
9	1.075766066087	1.069448852721
10	1.068297188921	1.063666938404

*See Definition 5.1 and the remarks above for a description of the constants β_q and γ_q.

THEOREM 5.1

Suppose that $f \in LC^q[a, b; M, \alpha]$ (see Section 2.2); $\zeta \in (a, b)$; $f^{(q-1)}(\zeta) = 0$; and $f^{(q)}(\zeta) \neq 0$. If $x_0, \ldots, x_q$ are (distinct and) sufficiently close to ζ, then a sequence (x_n) is uniquely defined by (1.1), and $x_n \to \zeta$ with weak order at least $\beta_{q,\alpha}$, the positive real root of $x^{q+1} = x + \alpha$.

Remark

If $\delta_0 = \max\limits_{i=0,\ldots,q} |x_i - \zeta|$ then, from Theorem 4.1, $x_0, \ldots, x_q$ are "sufficiently close" to ζ if $\delta_0 \leq \zeta - a$, $\delta_0 \leq b - \zeta$, and

$$3M\delta_0^\alpha < |f^{(q)}(\zeta)|. \tag{5.4}$$

If these conditions are satisfied, then an upper bound on $|x_n - \zeta|$ follows from equation (5.10) below.

Proof of Theorem 5.1

For $n \geq 0$, let

$$\delta_n = \max\limits_{i=0,\ldots,q} |x_{n+i} - \zeta|. \tag{5.5}$$

Suppose that $x_0, \ldots, x_q$ are so close to ζ that the conditions mentioned in the remark above are satisfied. Then Theorem 4.1 shows that (δ_n) is monotonic decreasing to zero, and

$$\delta_{n+q+1} \leq \frac{3M}{|f^{(q)}(\zeta)|} \delta_n^\alpha \delta_{n+1}. \tag{5.6}$$

If eventually $\delta_n = 0$, then the result follows immediately: by our definition, $x_n \to \zeta$ with weak order ∞. Hence, suppose that $\delta_n \neq 0$ for all $n \geq 0$. Let

$$\lambda_n = -\log\left(\delta_n \left|\frac{3M}{f^{(q)}(\zeta)}\right|^{1/\alpha}\right) \tag{5.7}$$

(not the same λ_n as in Theorem 4.1). From condition (5.4) and the fact that (δ_n) is monotonic decreasing, $0 < \lambda_0 \leq \lambda_1 \leq \lambda_2 \leq \cdots$, and, from equation (5.6),

$$\lambda_{n+q+1} \geq \lambda_{n+1} + \alpha\lambda_n. \tag{5.8}$$

Since $\beta_{q,\alpha} > 1$, we have

$$\lambda_n \geq \lambda_0 \beta_{q,\alpha}^{n-q} \tag{5.9}$$

for $n = 0, \ldots, q$. Thus, from (5.8) and the definition of $\beta_{q,\alpha}$, the inequality (5.9) holds for all $n \geq 0$, by induction on n. Hence, for all $n \geq 0$,

$$-\log|x_n - \zeta| \geq -\log\delta_n \geq \lambda_0 \beta_{q,\alpha}^{n-q} + \frac{1}{\alpha}\log\left|\frac{3M}{f^{(q)}(\zeta)}\right|. \tag{5.10}$$

Since $\lambda_0 > 0$ and $\beta_{q,\alpha} > 1$, equation (5.10) shows that

$$\liminf\limits_{n\to\infty} (-\log|x_n - \zeta|)^{1/n} \geq \beta_{q,\alpha}, \tag{5.11}$$

which completes the proof.

Note that, in the important case $\alpha = 1$, there is a simple proof of Theorem 5.1 which does not depend on Theorems 2.5.1 and 4.1. This proof shows that, instead of (5.4), the condition

$$3M\delta_0 < 2|f^{(q)}(\zeta)| \tag{5.12}$$

is sufficient. The idea is this: by applying Rolle's Theorem $q - 1$ times, we see that $P_n^{(q-1)}(x)$ coincides with $f^{(q-1)}(x)$ at points ξ_n and ξ_n' such that $|\xi_n - \zeta| \leq \delta_n$ and $|\xi_n' - \zeta| \leq \delta_n' = $ the second largest of $|x_n - \zeta|, \ldots,$ $|x_{n+q} - \zeta|$. Thus, from Lemma 2.4.1,

$$|P_n^{(q-1)}(\zeta)| \leq \frac{1}{2}M\delta_n\delta_n'. \tag{5.13}$$

On the other hand, equations (1.1) and (3.3) show that

$$x_{n+q+1} = \zeta - \frac{P_n^{(q-1)}(\zeta)}{q!\,f[x_n, \ldots, x_{n+q}]}, \tag{5.14}$$

so we can bound $|x_{n+q+1} - \zeta|$, and then the result follows in much the same way as above.

Section 6
THE EXACT ORDER OF CONVERGENCE

Theorem 5.1 gives conditions under which $x_n \to \zeta$ with weak order at least β_q. It is natural to ask if the order is exactly β_q. In general this is true, but some conditions are necessary to ensure that the rate of convergence is not too fast: for example, the successive linear interpolation process $(q = 1)$ converges to a simple zero ζ with weak order at least $2 (> \beta_1 = 1.618 \ldots)$ if it happens that $f''(\zeta) = 0$, for then linear interpolation is more accurate than would normally be expected. Theorem 6.1 gives sufficient conditions for the order to be exactly β_q. Apart from the condition $f^{(q+1)}(\zeta) \neq 0$, it is necessary to impose some conditions on the initial points $x_0, \ldots, x_q$. (These extra conditions are superfluous if $q = 1$: see Section 7.)

Before proving Theorem 6.1, we need two lemmas. Lemma 6.2 is concerned with the solution of a certain difference equation, and is closely related to Theorem 12.1 of Ostrowski (1966). The lemma could easily be generalized, but we only need the result stated. Lemma 6.1 gives a recurrence relation for the error $x_n - \zeta$. Special cases of this lemma have been given by Ostrowski (1966) and Jarratt (1967, 1968). Ostrowski essentially gives the case $q = 1$, and Jarratt gives weaker results for $q = 2$ and $q = 3$. (Our bound on the remainder R_n is sharper than Jarratt's, and we do not assume that f is analytic.) In Section 8, we show how the result of Lemma 6.1 may be used to accelerate convergence of the sequence (x_n).

LEMMA 6.1

Suppose that $f \in C^{q+1}[a, b]$; $\zeta \in [a, b]$; $f^{(q-1)}(\zeta) = 0$; $f^{(q)}(\zeta) \neq 0$; $x_n, \ldots, x_{n+q}$ are (distinct) points in $[a, b]$; and x_{n+q+1} satisfies equation (1.1). Let δ_n be the largest of $|x_n - \zeta|, \ldots, |x_{n+q} - \zeta|$; and δ'_n the second largest. Then

$$x_{n+q+1} - \zeta = \frac{f^{(q+1)}(\zeta)}{q(q+1)f^{(q)}(\zeta)} \sum_{0 \leq i < j \leq q} (x_{n+i} - \zeta)(x_{n+j} - \zeta) + R_n, \qquad (6.1)$$

where

$$R_n = O\{\delta_n \, \delta'_n [\delta_n + w(f^{(q+1)}; \delta_n)]\} \qquad (6.2)$$

as $\delta_n \to 0$.

Proof

Without loss of generality, assume that $n = 0$ and $\zeta = 0$. Rearrange $x_0, \ldots, x_q$, if necessary, so that $|x_0| \leq |x_1| \leq \cdots \leq |x_q|$. From Lemma 3.1,

$$q \, x_{q+1} f[x_0, \ldots, x_q] = \left(\sum_{i=0}^{q-1} x_i \right) f[x_0, \ldots, x_q] - f[x_0, \ldots, x_{q-1}]. \qquad (6.3)$$

Thus, as $f^{(q-1)}(0) = 0 \neq f^{(q)}(0)$, Corollary 2.5.1 gives

$$q x_{q+1} \frac{f^{(q)}(0)}{q!}(1 + r_1) = \left(\sum_{i=0}^{q-1} x_i \right) \left[\frac{f^{(q)}(0)}{q!} + \left(\sum_{i=0}^{q} x_i \right) \frac{f^{(q+1)}(0)}{(q+1)!} + r_2 \right]$$
$$- \left[\left(\sum_{i=0}^{q-1} x_i \right) \frac{f^{(q)}(0)}{q!} + \left(\sum_{0 \leq i \leq j < q} x_i x_j \right) \frac{f^{(q+1)}(0)}{(q+1)!} + r_3 \right], \qquad (6.4)$$

where

$$|r_1| \leq \frac{w(f^{(q)}; \delta_0)}{|f^{(q)}(0)|} = O(\delta_0), \qquad (6.5)$$

$$|r_2| \leq \frac{\delta_0 w(f^{(q+1)}; \delta_0)}{q!} = O(\delta_0 w(f^{(q+1)}; \delta_0)), \qquad (6.6)$$

and

$$|r_3| \leq \frac{\delta_0'^2 w(f^{(q+1)}; \delta_0')}{2(q-1)!} = O(\delta_0'^2 w(f^{(q+1)}; \delta_0')) \qquad (6.7)$$

as $\delta_0 \to 0$.

The right side of (6.4) is just

$$\left(\sum_{0 \leq i < j \leq q} x_i x_j \right) \frac{f^{(q+1)}(0)}{(q+1)!} + r_4, \qquad (6.8)$$

where

$$|r_4| \leq q\delta_0' |r_2| + |r_3| = O(\delta_0 \, \delta'_0 w(f^{(q+1)}; \delta_0)) \qquad (6.9)$$

as $\delta_0 \to 0$, so the result follows.

Remarks

From the bounds on $r_1, \ldots, r_4$, it is easy to derive an explicit bound on $|R_n|$ for sufficiently small δ_n. For our purposes, though, the relation (6.2) is adequate. A simple corollary of (6.2) is that, if $f^{(q+1)} \in Lip_M \alpha$, then

$$R_n = O(\delta_n^{1+\alpha} \delta_n') \tag{6.10}$$

as $\delta_n \to 0$.

LEMMA 6.2

Suppose that $\lambda_n \to +\infty$ as $n \to \infty$ and, for $n \geq 0$,

$$\lambda_{n+q+1} - \lambda_{n+1} - \lambda_n = k_n, \tag{6.11}$$

where

$$k_n = O(s^n) \tag{6.12}$$

as $n \to \infty$, s a constant. If $\gamma_q < s < \beta_q$ then

$$\lambda_n = c\beta_q^n + O(s^n) \tag{6.13}$$

as $n \to \infty$, and if $k_n = o(s^n)$ as $n \to \infty$ then

$$\lambda_n = c\beta_q^n + o(s^n) \tag{6.14}$$

as $n \to \infty$. If $0 \leq s < \gamma_q$ then

$$\lambda_n = c\beta_q^n + O(n^v\gamma_q^n) \tag{6.15}$$

as $n \to \infty$, where

$$v = \begin{cases} 0 & \text{if } q = 1, \\ 1 & \text{if } q > 1, \end{cases} \tag{6.16}$$

and c is a nonnegative constant.

Proof

The restriction $|u_2| < 1$ in Theorem 12.1 of Ostrowski (1966) is unnecessary, for we can choose any λ with $|u_2| < \lambda < |u_1|$, and consider λ_n/λ^n instead of λ_n in Ostrowski's proof. Thus, in view of the remarks after Definition 5.1, (6.13) and (6.15) follow from Ostrowski's Theorem 12.1. (6.14) does not follow directly in the same way, but the proof of Ostrowski's Theorem 12.1 goes through, assuming $k_n = o(s^n)$ instead of $k_n = O(s^n)$, and giving a result from which (6.14) follows. The only difficulty is in proving the modified form of Ostrowski's Lemma 12.1, but this follows from the Toeplitz lemma: if $k_n \to 0$, $|\xi| < 1$, and $z_n = k_n + k_{n-1}\xi + \cdots + k_0\xi^n$, then $z_n \to 0$ as $n \to \infty$ (see Ortega and Rheinboldt (1970), pg. 399).

THEOREM 6.1

Let $f \in C^{q+1}[a,b]$; $\zeta \in (a,b)$; $f^{(q-1)}(\zeta) = 0$; $f^{(q)}(\zeta) \neq 0$; and $f^{(q+1)}(\zeta) \neq 0$. Suppose that $|x_0 - \zeta|$ is sufficiently small, that

$$|x_{i-1} - \zeta| \geq 4|x_i - \zeta| \tag{6.17}$$

for $i = 1, 2, \ldots, q,$ and that

$$|x_q - \zeta| \geq 6\,|K(x_0 - \zeta)(x_1 - \zeta)| > 0, \qquad (6.18)$$

where

$$K = \frac{f^{(q+1)}(\zeta)}{q(q+1)f^{(q)}(\zeta)}. \qquad (6.19)$$

Then a sequence (x_n) is uniquely defined by (1.1), and $x_n \to \zeta$ with weak order exactly β_q. In fact, if $q = 1$ or 2 then $x_n \to \zeta$ with strong order β_q and asymptotic constant $|K|^{\beta_q - 1}$, and if $q \geq 3$ then

$$-\log|x_n - \zeta| = c\beta_q^n + O(n\gamma_q^n) \qquad (6.20)$$

as $n \to \infty$, for some positive constant c.

Remarks

Condition (6.17) ensures that $x_0, \ldots, x_q$ approach ζ sufficiently fast, while (6.18) makes sure they do not approach ζ too fast. These conditions could be weakened, but Theorem 7.1 shows that some such conditions are necessary if $q \geq 2$. If $q = 1$ then the conditions are superfluous: see Corollary 7.1.

Equation (6.20) implies that (2.2) holds with $\rho = \beta_q$, but (2.1) does not necessarily hold, for $\gamma_q > 1$ if $q \geq 3$.

Proof of Theorem 6.1
Let

$$y_n = |K(x_n - \zeta)|. \qquad (6.21)$$

From the assumptions (6.17) and (6.18) we have, at least for $n = 0$,

$$y_{n+i-1} \geq 4y_{n+i} \qquad (6.22)$$

for $i = 1, 2, \ldots, q,$ and

$$y_{n+q} \geq 6y_n y_{n+1} > 0. \qquad (6.23)$$

We shall show that (6.22) and (6.23) hold for all $n \geq 0$. Suppose, as inductive hypothesis, that they hold for all $n \leq m$. Then, by taking $|x_0 - \zeta|$ sufficiently small (independent of m), we may suppose that the remainder R_n of Lemma 6.1 satisfies

$$|KR_n| \leq \frac{1}{13} y_n y_{n+1} \qquad (6.24)$$

for all $n \leq m$. Thus, from Lemma 6.1,

$$y_{m+q+1} \leq y_m y_{m+1}\left[\left(1 + \frac{1}{4} + \frac{2}{4^2} + \frac{2}{4^3} + \frac{3}{4^4} + \cdots\right) + \frac{1}{13}\right]$$
$$\leq \frac{3}{2} y_m y_{m+1}. \qquad (6.25)$$

From (6.23) with $n = m$, this gives

$$y_{m+q} \geq 4y_{m+q+1}. \tag{6.26}$$

Similarly,

$$y_{m+q+1} \geq y_m y_{m+1}\left[\left(1 - \frac{1}{4} - \frac{2}{4^2} - \frac{2}{4^3} - \frac{3}{4^4} - \cdots\right) - \frac{1}{13}\right]$$

$$\geq \frac{1}{2} y_m y_{m+1} \tag{6.27}$$

$$\geq 6y_{m+1} y_{m+2}. \tag{6.28}$$

Also, from (6.27), $y_{m+q+1} > 0$, so the right side of (6.28) is positive. From (6.26) and (6.28), we see that (6.22) and (6.23) hold for $n = m + 1$, so they hold for all $n \geq 0$, by induction. Thus (6.25) and (6.27) hold for all $m \geq 0$.

Let

$$\lambda_n = -\log y_n \tag{6.29}$$

and

$$k_n = \lambda_{n+q+1} - \lambda_{n+1} - \lambda_n. \tag{6.30}$$

From (6.25) and (6.27),

$$|k_n| \leq \log 2, \tag{6.31}$$

so we may apply Lemma 6.2 with $s = 1$. If $q \geq 3$ then $\gamma_q > 1$, so

$$\lambda_n = c\beta_q^n + O(n\gamma_q^n) \tag{6.32}$$

as $n \to \infty$. From Theorem 5.1, $c > 0$, so the result for $q \geq 3$ follows.

If $q = 1$ or 2 then $\gamma_q < 1$, so

$$\lambda_n = c\beta_q^n + O(1) \tag{6.33}$$

as $n \to \infty$. From (6.29), (6.30), (6.33) and Lemma 6.1, we now see that

$$k_n = o(1) \tag{6.34}$$

as $n \to \infty$, so, by equation (6.14) with $s = 1$,

$$\lambda_n = c\beta_q^n + o(1) \tag{6.35}$$

as $n \to \infty$. Thus, there exists

$$\lim_{n \to \infty} \frac{y_{n+1}}{y_n^{\beta_q}} = 1, \tag{6.36}$$

so the result follows from equation (6.21). Note that, if $f^{(q+1)} \in Lip_M \alpha$ for any M and $\alpha > 0$, then (6.34) may be replaced by $k_n = o(s^n)$ for any $s > 0$, so (6.15) holds, and

$$\frac{|x_{n+1} - \zeta|}{|x_n - \zeta|^{\beta_q}} = |K|^{\beta_q - 1} + O(n^{q-1}\gamma_q^n) \tag{6.37}$$

as $n \to \infty$.

Section 7
STRONGER RESULTS FOR $q = 1$ AND 2

In this section we restrict our attention to the two cases of greatest practical interest: $q = 1$ (successive linear interpolation) and $q = 2$ (successive parabolic interpolation for finding an extreme point). Corollary 7.1 shows that the conditions (6.17) and (6.18) of Theorem 6.1 are unnecessary if $q = 1$.

COROLLARY 7.1

Suppose that $q = 1$; $f \in C^2[a, b]$; $\zeta \in (a, b)$; $f(\zeta) = 0$; $f'(\zeta) \neq 0$; and $f''(\zeta) \neq 0$. If x_0, x_1 and ζ are distinct and sufficiently close together, then a sequence (x_n) is uniquely defined by (1.1), and $x_n \to \zeta$ with strong order $\beta_1 = \frac{1}{2}(1 + \sqrt{5})$ and asymptotic constant $|f''(\zeta)/(2f'(\zeta))|^{\beta_1 - 1}$ as $n \to \infty$.

Proof

From Lemma 6.1,

$$x_2 - \zeta = \frac{f''(\zeta)}{2f'(\zeta)}(x_0 - \zeta)(x_1 - \zeta)(1 + o(1)) \tag{7.1}$$

as $\max(|x_0 - \zeta|, |x_1 - \zeta|) \to 0$. Thus, Theorem 6.1 is applicable to the sequence (x'_n), where $x'_n = x_{n+1}$, provided x_0 and x_1 are sufficiently close to ζ.

Remarks

Ostrowski (1966) gives Corollary 7.1 with the stronger assumption that $f \in C^3[a, b]$. He also shows that, if $f \in C^3[a, b]$ and the conditions of Corollary 7.1 are satisfied, then

$$\frac{|x_{n+1} - \zeta|}{|x_n - \zeta|^{\beta_1}} = \left|\frac{f''(\zeta)}{2f'(\zeta)}\right|^{\beta_1 - 1} + O(\gamma_1^n) \tag{7.2}$$

as $n \to \infty$. As we remarked at the end of the proof of Theorem 6.1, the relation (7.2) holds provided that $f \in LC^2[a, b; M, \alpha]$ for some M and α (see equation (6.37)). For an even weaker condition, see (7.7) and (7.8) below.

The following theorem removes the rather artificial restrictions (6.17) and (6.18) of Theorem 6.1, if $f^{(q+1)}$ is Lipschitz continuous and $q = 1$ or 2. The proof does not extend to $q \geq 3$ because it depends on the assumption that $\gamma_q < 1$, which is only true for $q = 1$ and $q = 2$ (see Table 5.1).

THEOREM 7.1

Suppose that $q = 1$ or 2; $f \in LC^{(q+1)}[a, b; M]$; $\zeta \in (a, b)$; $f^{(q-1)}(\zeta) = 0$; and $f^{(q)}(\zeta) \neq 0$. If $x_0, \ldots, x_q$ are (distinct and) sufficiently close to ζ, then a sequence (x_n) is uniquely defined by (1.1), and either

1. $x_n \to \zeta$ with strong order β_q and asymptotic constant

$$\left| \frac{f^{(q+1)}(\zeta)}{q(q+1)f^{(q)}(\zeta)} \right|^{\beta_q - 1}, \text{ in fact}$$

$$\frac{|x_{n+1} - \zeta|}{|x_n - \zeta|^{\beta_q}} = \left| \frac{f^{(q+1)}(\zeta)}{q(q+1)f^{(q)}(\zeta)} \right|^{\beta_q - 1} + O(n^{q-1}\gamma_q^n) \tag{7.3}$$

as $n \to \infty$ (recall that $\beta_1 \simeq 1.618$, $\beta_2 \simeq 1.325$, $\gamma_1 \simeq 0.618$, and $\gamma_2 \simeq 0.869$); or

2. $x_n \to \zeta$ with weak order at least 2 if $q = 1$, or at least

$$[(3 + \sqrt{5})/2]^{1/3} \simeq 1.378 \qquad \text{if } q = 2.$$

Remarks

If $q = 1$ then, by Corollary 7.1, case 2 of Theorem 7.1 is possible only if $f''(\zeta) = 0$ (or if one of x_0 and x_1 coincides with ζ, when the weak order is ∞). If $q = 2$ then case 2 is possible, although unlikely, even if $f^{(3)}(\zeta) \neq 0$ and $x_n \neq \zeta$ for all n. All that is necessary is that the terms in relation (7.28) repeatedly nearly cancel out. Jarratt (1967) and Kowalik and Osborne (1968) assume that such cancellation will eventually die out, so the order will be β_2. The conditions (6.17) and (6.18) are sufficient for this to be true, but without some such conditions there is a remote possibility that cancellation will continue indefinitely. For example, with $f(x) = 2x^3 + x^2$, there are starting values x_0, x_1 and x_2 such that

$$\left. \begin{array}{l} x_{2n} \sim \exp(-2^n) \\[4pt] x_{2n+1} \sim -\exp(-2^n), \end{array} \right\} \tag{7.4}$$

and

so $x_n \to \zeta = 0$ with weak order $\sqrt{2}$. Similarly, if

$$\gamma = \tfrac{1}{2}(3 + \sqrt{5}) = 2.618 \ldots, \tag{7.5}$$

then there are starting values such that

$$\left. \begin{array}{l} x_{3n} \sim \exp(-\gamma^n), \\[4pt] x_{3n+1} \sim \exp(-(\gamma - 1)\gamma^n), \\[4pt] x_{3n+2} \sim -\exp(-(\gamma - 1)\gamma^{n+1}), \end{array} \right\} \tag{7.6}$$

and

so $x_n \to 0$ with weak order $\gamma^{1/3} = 1.378 \ldots$. The proof is omitted, but the reader may easily verify that (7.4) and (7.6) are compatible with Lemma 7.3 below (this depends on the relation $2\gamma - 1 = \gamma(\gamma - 1)$).

For the sake of simplicity, we have not stated Theorem 7.1 in the sharpest possible form. If $f^{(q+1)}(\zeta) = 0$, then $x_n \to \zeta$ with weak order at least $\beta_{q,1+\alpha} > \beta_q$, provided that $f^{(q+1)} \in Lip_M \alpha$ for some M and $\alpha > 0$. If $f^{(q+1)}(\zeta) \neq 0$, then the theorem holds provided that $f \in C^{q+1}[a, b]$. Equation (7.3) may no longer hold, but if there is an $\epsilon > 0$ such that

$$w(f^{(q+1)}; \delta) = O(|\log \delta|^{-\epsilon/q}) \tag{7.7}$$

as $\delta \to 0$, then

$$\frac{|x_{n+1} - \zeta|}{|x_n - \zeta|^{\beta_q}} - \left|\frac{f^{(q+1)}(\zeta)}{q(q+1)f^{(q)}(\zeta)}\right|^{\beta_q - 1} = \begin{cases} O(n^{q-1}\gamma_q^n) & \text{if} \quad \epsilon > 1, \\ O(n^q\gamma_q^n) & \text{if} \quad \epsilon = 1, \\ O(\gamma_q^{n\epsilon}) & \text{if} \quad \epsilon < 1, \end{cases} \quad (7.8)$$

as $n \to \infty$. (A condition like (7.7) occurs in some variants of Jackson's theorem: see Meinardus (1967).)

Before proving Theorem 7.1, we need three rather technical lemmas.

LEMMA 7.1

Suppose that, for $n \geq 0$,

$$x_{n+3} = x_n x_{n+1} + x_{n+1} x_{n+2} + x_n x_{n+2} + m_n \delta_n^2 \delta_n', \quad (7.9)$$

where δ_n is the largest of $|x_n|$, $|x_{n+1}|$ and $|x_{n+2}|$, and δ_n' is the second largest. If there is a positive constant L such that

$$\frac{1}{15L} \geq |x_0| \geq 3|x_1| \geq 9|x_2| \geq 27|x_3|,$$

and

$$|m_n| \leq L \quad (7.10)$$

for all $n \geq 0$, then $|x_n| \geq 3|x_{n+1}|$ for all $n \geq 0$.

Proof

As in the proof of Theorem 6.1, it follows by induction on n that

$$|x_{n+3}| \geq \frac{22}{45}|x_n x_{n+1}| \geq \frac{22}{5}|x_{n+1} x_{n+2}| \geq 3|x_{n+4}| \quad (7.11)$$

for all $n \geq 0$.

LEMMA 7.2

If the conditions of Lemma 7.1 are satisfied, then either $x_n = 0$ for all sufficiently large n, or

$$\frac{|x_{n+1}|}{|x_n|^{\beta_2}} = 1 + O(n\gamma_2^n)$$

as $n \to \infty$.

Proof

If $x_n \neq 0$ for infinitely many n then, by Lemma 7.1, $x_n \neq 0$ for all $n \geq 0$. If this is so, define $\lambda_n = -\log|x_n|$ and $k_n = \lambda_{n+3} - \lambda_{n+1} - \lambda_n$. From equation (7.11), k_n is bounded, so Lemma 6.2 with $s = 1$ gives $\lambda_n = c\beta_2^n + O(1)$ as $n \to \infty$. By Lemma 7.1, $\lambda_n \to +\infty$, so $c > 0$. Thus, from (7.9),

$$k_n = O(\exp\{-c(\beta_2 - 1)\beta_2^{n+1}\}) \quad (7.12)$$

as $n \to \infty$. (This is not necessarily true in the proof of Theorem 6.1.) Now,

Lemma 6.2 with $s < \gamma_2$ gives

$$\lambda_n = c\beta_2^n + O(n\gamma_2^n) \tag{7.13}$$

as $n \to \infty$, and the result follows from the definition of λ_n.

LEMMA 7.3

Suppose that (7.9) and (7.10) hold. There are constants K and N (depending on L) such that if, for some $n \geq N$,

$$\frac{1}{n} \geq |x_n| \geq n|x_{n+2}| \tag{7.14}$$

and

$$\frac{1}{n} \geq |x_{n+1}| \geq n|x_{n+2}|, \tag{7.15}$$

then

$$x_{n+3} = x_n x_{n+1}(1 + v_{1,n}), \tag{7.16}$$

$$x_{n+4} = x_n x_{n+1}^2(1 + v_{2,n}) + x_{n+1} x_{n+2}(1 + v_{3,n}), \tag{7.17}$$

$$x_{n+5} = x_n^2 x_{n+1}^3(1 + v_{4,n}) + x_n x_{n+1} x_{n+2}(1 + v_{5,n}), \tag{7.18}$$

and

$$x_{n+6} = x_n^2 x_{n+1}^3(1 + v_{6,n}) + x_n x_{n+1}^2 x_{n+2}(1 + v_{7,n}), \tag{7.19}$$

where

$$|v_{i,n}| \leq \frac{K}{n} \tag{7.20}$$

for $i = 1, \ldots, 7$.

Proof

The lemma follows by repeated use of the recurrence relation (7.9) and the inequalities (7.10), (7.14), and (7.15).

Proof of Theorem 7.1

Without loss of generality assume that $\zeta = 0$. First suppose that $q = 1$. If $f''(0) \neq 0$ then the theorem holds, by Corollary 7.1. If $f''(0) = 0$ then, by Lemma 6.1,

$$x_{n+2} = O(\delta_n^2 \delta_n') \tag{7.21}$$

as $\delta_n \to 0$, where δ_n and δ_n' are as in Lemma 6.1. If x_0 and x_1 are sufficiently small, equation (7.21) implies that

$$\delta_n = |x_n| \tag{7.22}$$

and

$$\delta_n' = |x_{n+1}| \tag{7.23}$$

for all $n \geq 1$. Thus $x_n \to 0$ as $n \to \infty$, and

$$|x_{n+2}| \leq A^2 |x_n^2 x_{n+1}| \qquad (7.24)$$

for all $n \geq 0$, where A is some positive constant. If some $x_n = 0$ then $x_{n+1} = x_{n+2} = \cdots = 0$, and we are finished (weak order ∞). Otherwise, there is no loss of generality in assuming that

$$A |x_n| \leq \exp(-2^n) \qquad (7.25)$$

for $n = 0$ and $n = 1$. From (7.24), equation (7.25) holds for all $n \geq 0$, by induction on n. Thus, the weak order of convergence is at least 2, and the proof for $q = 1$ is complete.

From now on suppose that $q = 2$. By Lemma 6.1,

$$x_{n+3} = \frac{f^{(3)}(0)}{6f''(0)}(x_n x_{n+1} + x_{n+1} x_{n+2} + x_n x_{n+2}) + O(\delta_n^2 \delta_n') \qquad (7.26)$$

as $n \to \infty$. If $f^{(3)}(0) = 0$ then the weak order of convergence is at least $\beta_{2,2}$, the positive real root of $x^3 = x + 2$, by a proof like that above for $q = 1$, and the theorem holds as $\beta_{2,2} = 1.52 \ldots > 1.38$.

If $f^{(3)}(0) \neq 0$, then we may as well suppose that

$$\frac{f^{(3)}(0)}{6f''(0)} = 1 \qquad (7.27)$$

by a change of scale, as in the proof of Theorem 6.1. Thus, we must study the interesting recurrence relation

$$x_{n+3} = x_n x_{n+1} + x_{n+1} x_{n+2} + x_n x_{n+2} + O(\delta_n^2 \delta_n'), \qquad (7.28)$$

and, by Theorem 5.1, we can assume that $x_n \to 0$ with weak order at least β_2.

First suppose that there is an infinite sequence $N = (n_0, n_1, \ldots)$ with the property that, for every $i \geq 0$ and $n = n_i$, either

1. $$n_{i+1} = n + 2 \qquad (7.29)$$

and

$$4n |x_n x_{n+1}^2| \leq |x_{n+2}| \leq 2 |x_n x_{n+1}|, \qquad (7.30)$$

or

2. $$n_{i+1} = n + 3 \qquad (7.31)$$

and

$$|x_{n+2}| < 4n |x_n x_{n+1}^2|. \qquad (7.32)$$

If either (7.30) or (7.32) holds, then Lemma 7.3 is applicable for all sufficiently large $n = n_i$ in the sequence N. To avoid confusion with subscripts, write m for n_{i+1} (so $m = n + 2$ or $n + 3$). If $n = n_i$ is sufficiently large, and (7.29) and (7.30) hold, then

$$|x_m| \leq 2 |x_n x_{n+1}| \qquad (7.33)$$

and, by Lemma 7.3,

$$|x_{m+1}| \leq 2|x_n x_{n+1}|. \tag{7.34}$$

If (7.31) and (7.32) hold then, similarly,

$$|x_m| \leq 2|x_n x_{n+1}| \tag{7.35}$$

and

$$|x_{m+1}| \leq 4|x_n x_{n+1}^2|. \tag{7.36}$$

Let

$$y_n = 2|x_n|. \tag{7.37}$$

After a fixed $n = n_i$ in N, suppose that the next $r \geq 1$ elements of N satisfy (7.31), and then the next $s \geq 1$ satisfy (7.29). Then repeated use of the inequalities (7.33) to (7.36) gives

$$\max(y_{n+3r+2s}, y_{n+3r+2s+1}) \leq \max(y_n, y_{n+1})^{\varphi(r,s)}, \tag{7.38}$$

where

$$\varphi(r, s) = 2^{s-1}\left[\left(\frac{\sqrt{5}+2}{\sqrt{5}}\right)\left(\frac{3+\sqrt{5}}{2}\right)^r + \left(\frac{\sqrt{5}-2}{\sqrt{5}}\right)\left(\frac{3-\sqrt{5}}{2}\right)^r\right]. \tag{7.39}$$

Let

$$\psi(r, s) = \varphi(r, s)^{1/(3r+2s)}. \tag{7.40}$$

For fixed $s \geq 1$, $\psi(r, s)$ is a decreasing function of r, with limit

$$c = \left(\frac{3+\sqrt{5}}{2}\right)^{1/3} = \inf_{r,s\geq 1} \psi(r, s) \tag{7.41}$$

as $r \to \infty$. Thus, $x_n \to 0$ with weak order at least c, so case 2 of the theorem holds.

Now suppose that there is no such infinite sequence N. By the superlinear convergence of (x_n), Lemma 7.3 is applicable for infinitely many n. Choose such an n (sufficiently large). There are only three possibilities:

1. Equation (7.30) holds;
2. Equation (7.32) holds; or
3. Neither (7.30) nor (7.32) holds, so

$$|x_{n+2}| > 2|x_n x_{n+1}|. \tag{7.42}$$

In the first case, Lemma 7.3 shows that we can replace n by $n + 2$, and continue with one of the three cases (it is crucial to note that Lemma 7.3 is still applicable). In the second case, Lemma 7.3 shows that we can replace n by $n + 3$ and continue. Since no infinite sequence N with the above properties exists, the third case must eventually arise. Then, from (7.42) and Lemma 7.3, we see that Lemma 7.2 is applicable to the sequence (x'_m), where $x'_m = x_{m+n+3}$. By Lemma 7.2, (x'_m) converges with strong order β_2 and asymptotic constant 1, and hence so does (x_n). In view of the assumption (7.27), this completes the proof.

Section 8
ACCELERATING CONVERGENCE

If a very accurate solution is required, and high-precision evaluations of f are expensive, then it may be worthwhile to try to increase the order of convergence of the successive approximations by some acceleration technique. For example, we can use Lemma 6.1 to improve the current approximation at each step of the iterative process. Jarratt (1967) suggests one way of doing this if $q = 2$, but the method which we are about to describe seems easier to justify (see Theorem 8.1), and applies for any $q \geq 1$.

Suppose that $x_0, \ldots, x_{q+1}$ are approximations to a simple zero ζ of $f^{(q-1)}$. For example, they could be the last $q + 2$ approximations generated by the successive interpolation process discussed above. We may define $x_{q+2}, x_{q+3}, \ldots$ in the following way: if $n \geq 1$ and $x_0, \ldots, x_{n+q}$ are already defined, let $P_n = IP(f; x_n, \ldots, x_{n+q})$, and choose y_n such that

$$P_n^{(q-1)}(y_n) = 0. \tag{8.1}$$

I.e., y_n is just the next approximation generated by our usual interpolation process. From Lemma 3.1, y_n is given explicitly by

$$y_n = \frac{1}{q}\left(\sum_{i=1}^{q} x_{n+i} - \frac{f[x_{n+1}, \ldots, x_{n+q}]}{f[x_n, \ldots, x_{n+q}]}\right). \tag{8.2}$$

Instead of taking y_n as the next approximation x_{n+q+1}, we use Lemma 6.1 to compute a correction to y_n, and take the corrected value as the next approximation. Formally, we define x_{n+q+1} by

$$x_{n+q+1} = y_n - \left(\frac{f[x_{n-1}, \ldots, x_{n+q}]}{q f[x_n, \ldots, x_{n+q}]}\right)s_n, \tag{8.3}$$

where

$$s_n = \sum_{0 \leq i < j \leq q} (x_{n+i} - y_n)(x_{n+j} - y_n). \tag{8.4}$$

For a justification of equations (8.3) and (8.4), see the proof of Theorem 8.1 below. This theorem shows that, under suitable conditions, the sequence (x_n) is well-defined, and $x_n \to \zeta$ with weak order appreciably greater than β_q, which is the usual order of convergence of the unaccelerated process (see Sections 5 to 7). Note that there is very little extra work involved in computing x_{n+q+1} from equations (8.3) and (8.4) if y_n is computed via (8.2), for $f[x_n, \ldots, x_{n+q}]$ and $f[x_{n-1}, \ldots, x_{n+q-1}]$ will already be known, except at the first iteration.

Before stating Theorem 8.1, we define some constants β_q' which take the place of the constants β_q (Definition 5.1) if the accelerated process is used.

DEFINITION 8.1

For $q \geq 1$, β_q' is the positive real root of

$$x^{q+2} = x^2 + x + 1. \tag{8.5}$$

Remarks

It is easy to see that $\beta_q' > \beta_q$ and, corresponding to the bound (5.2), we have

$$3^{1/(q+1)} < \beta_q' < 3^{1/q}. \tag{8.6}$$

If $x_n \to \zeta$ with weak order $\beta > 1$ then, by the definition of order (see Section 2), for any $\epsilon > 0$ we eventually have

$$-\log|x_n - \zeta| \geq (\beta - \epsilon)^n. \tag{8.7}$$

Assuming that approximate equality holds in (8.7), the number of function evaluations required to reduce $|x_n - \zeta|$ below a very small positive tolerance is inversely proportional to $\log \beta$. Thus, the ratio $(\log \beta_q)/\log \beta_q'$ suggests how much we gain by using the accelerated process, rather than the unaccelerated process, if very high accuracy is required. From the bounds (5.2) and (8.6),

$$\lim_{q \to \infty} \frac{\log \beta_q}{\log \beta_q'} = \log_3 2 = 0.6309 \ldots, \tag{8.8}$$

so there is a 37 percent saving for large q. Of course, the only practical interest is in small values of q, and in Table 8.1 the values of β_q', β_q and $(\log \beta_q)/\log \beta_q'$ are given for $q = 1, 2, \ldots, 10$. The entries for β_q' are correctly rounded

TABLE 8.1   **The constants** β_q' **for** $q = 1(1)10$*

q	β_q'	β_q	$(\log \beta_q)/\log \beta_q'$
1	1.839286755214	1.6180	0.7897
2	1.465571231877	1.3247	0.7357
3	1.324717957245	1.2207	0.7093
4	1.249851588864	1.1673	0.6936
5	1.203216033518	1.1347	0.6832
6	1.171321856385	1.1128	0.6757
7	1.148113497353	1.0970	0.6702
8	1.130459571864	1.0851	0.6658
9	1.116575158368	1.0758	0.6623
10	1.105367322949	1.0683	0.6595

*See Definition 8.1, and the remarks above, for a description of the constants β_q' and the significance of the ratio $(\log \beta_q)/\log \beta_q'$.

to 12 decimal places, and the other entries are given to four places. (See Table 5.1 for the β_q to 12 places.) The table suggests that $\beta_3' = \beta_2$, and this is true, for $x^5 - x^2 - x - 1 = (x^3 - x - 1)(x^2 + 1)$.

THEOREM 8.1

Suppose that $f \in LC^{q+1}[a, b; M]$; $\zeta \in (a, b)$; $f^{(q-1)}(\zeta) = 0$; $f^{(q)}(\zeta) \neq 0$; and $x_0, \ldots, x_{q+1}$ are (distinct) points in $[a, b]$. If $x_0, \ldots, x_{q+1}$ are sufficiently close to ζ, then a sequence (x_n) is uniquely defined by equations (8.2) to (8.4), and $x_n \to \zeta$ with weak order at least β_q' (Definition 8.1) as $n \to \infty$.

Proof

For $n \geq 1$, let δ_n be the largest of $|x_n - \zeta|, \ldots, |x_{n+q} - \zeta|$; let δ'_n be the second-largest; and let

$$\hat{\delta}_n = \max(\delta_n, |x_{n-1} - \zeta|). \tag{8.9}$$

If y_n is defined by equation (8.2), then Lemma 6.1 shows that

$$y_n - \zeta = K \sum_{0 \leq i < j \leq q} (x_{n+i} - \zeta)(x_{n+j} - \zeta) + O(\delta_n^2 \delta'_n) \tag{8.10}$$

as $\delta_n \to 0$, where

$$K = \frac{f^{(q+1)}(\zeta)}{q(q+1)f^{(q)}(\zeta)}. \tag{8.11}$$

In particular, (8.10) implies that

$$y_n - \zeta = O(\delta_n \delta'_n) \tag{8.12}$$

as $\delta_n \to 0$. Thus, for $0 \leq i < j \leq q$,

$$(x_{n+i} - y_n)(x_{n+j} - y_n) = (x_{n+i} - \zeta)(x_{n+j} - \zeta) + O(\delta_n^2 \delta'_n) \tag{8.13}$$

as $\delta_n \to 0$.

If δ_n is sufficiently small then, since $f^{(q)}(\zeta) \neq 0$, we have $f[x_n, \ldots, x_{n+q}] \neq 0$ and, by Theorem 2.5.1,

$$\frac{f[x_{n-1}, \ldots, x_{n+q}]}{qf[x_n, \ldots, x_{n+q}]} = K + O(\hat{\delta}_n) \tag{8.14}$$

as $\hat{\delta}_n \to 0$.

If s_n is as in (8.4), then (8.13) and (8.14) give

$$\left(\frac{f[x_{n-1}, \ldots, x_{n+q}]}{qf[x_n, \ldots, x_{n+q}]}\right)s_n = K \sum_{0 \leq i < j \leq q} (x_{n+i} - \zeta)(x_{n+j} - \zeta) + O(\hat{\delta}_n \delta_n \delta'_n)$$
$$\tag{8.15}$$

as $\hat{\delta}_n \to 0$. Thus, from (8.3) and (8.10),

$$x_{n+q+1} - \zeta = O(\hat{\delta}_n \delta_n \delta'_n) \tag{8.16}$$

as $\hat{\delta}_n \to 0$. This shows that, provided $\hat{\delta}_1$ is sufficiently small, the sequence (x_n) is uniquely defined, lies in $[a, b]$, and $x_n \to \zeta$ as $n \to \infty$.

From equation (8.16), there is a positive constant A such that, for all $n \geq 1$,

$$|x_{n+q+1} - \zeta| \leq A^2 \hat{\delta}_n \delta_n \delta'_n. \tag{8.17}$$

Also, if $\hat{\delta}_1$ is sufficiently small, then

$$-\log(A|x_n - \zeta|) \geq \beta_q'^n \tag{8.18}$$

for $n = 0, \ldots, q+1$. From equation (8.17) and the definition of β'_q, we see that (8.18) holds for all $n \geq 0$, by induction on n. Thus

$$\liminf_{n \to \infty} (-\log|x_n - \zeta|)^{1/n} \geq \beta'_q, \tag{8.19}$$

i.e., the weak order of convergence is at least β'_q, so the proof is complete.

Section 9
SOME NUMERICAL EXAMPLES

To illustrate the theoretical results obtained in Sections 4 to 8, we give the following examples:

1. $q = 1$, $f(x) = x + x^2 + x^3$, $x_0 = 2$, $x_1 = 1$;
2. $q = 2$, $f(x) = 8 + 6x^2 + 4x^3 + 3x^4$, $x_0 = 2$, $x_1 = 1$, $x_2 = 0.5$;
3. $q = 3$, $f(x) = 1 + 40x + 10x^3 + 5x^4 + 3x^5$; $x_0 = 2$, $x_1 = 1$, $x_2 = 0.5$, $x_3 = 0.25$; and
4. $q = 4$, $f(x) = 1 + 2x + 40x^2 + 5x^4 + 2x^5 + x^6$, $x_0 = 2$, $x_1 = 1$, $x_2 = 0.5$, $x_3 = 0.25$, $x_4 = 0.125$.

In all these examples $\zeta = 0$, and the iterative process defined by (1.1) converges, even though the initial values are not very close to ζ. Apart from constant factors, the polynomials are obtained by differentiating the last one (Example 4) $4 - q$ times, so we are solving the same problem in four different ways.

Table 9.1 gives the sequences (x_n) produced by the successive interpolation process, for the functions and starting values given above. To illustrate the superlinear convergence, the entries are given until $|x_n| < 10^{-20}$, although such high precision would seldom be required in practical problems. The table also gives the sequences (x'_n) produced by the accelerated interpolation process described in Section 8, with starting values $x'_i = x_i$ for $i = 0, \ldots, q + 1$. As predicted by Theorem 8.1 and Table 8.1, the accelerated sequences converge appreciably faster than the unaccelerated ones.

To verify relations (8.12) and (8.16), the table gives

$$r_n = \frac{x_n}{x_{n-q}x_{n-q-1}} \qquad (9.1)$$

and

$$r'_n = \frac{x'_n}{x'_{n-q}x'_{n-q-1}x'_{n-q-2}} \qquad (9.2)$$

when they are defined. With a few exceptions near the beginning of some of the sequences, both $(|x_n|)$ and $(|x'_n|)$ are monotonic decreasing, so r_n and r'_n should be bounded. From Lemma 6.1, we expect that

$$\lim_{n \to \infty} r_n = \frac{f^{(q+1)}(\zeta)}{q(q + 1)f^{(q)}(\zeta)}, \qquad (9.3)$$

and this is just $2/[q(q + 1)]$ for our examples. Similarly, from the proof of Theorem 8.1, we expect that

$$\lim_{n \to \infty} r'_n = -\frac{f^{(q+2)}(\zeta)}{q(q + 1)(q + 2)f^{(q)}(\zeta)}, \qquad (9.4)$$

and this is just $-6/[q(q + 1)(q + 2)]$. The results support these predictions.

TABLE 9.1 Numerical results for $q = 1, 2, 3$ and 4

q	n	x_n	x_n'	r_n	r_n'
1	0	2.000	2.000		
	1	1.000	1.000		
	2	7.273'−1	7.273'−1	0.3636	
	3	3.980'−1	2.100'−1	0.5473	0.1444
	4	1.983'−1	4.389'−2	0.6851	0.2874
	5	6.727'−2	−1.846'−3	0.8523	−0.2755
	6	1.276'−2	1.221'−5	0.9568	−0.7178
	7	8.543'−4	1.035'−9	0.9949	−1.0455
	8	1.090'−5	2.350'−17	0.9998	−1.0066
	9	9.314'−9	−2.982'−31	1.0000	−1.0039
	10	1.015'−13		1.0000	
	11	9.457'−22		1.0000	
2	0	2.000	2.000		
	1	1.000	1.000		
	2	5.000'−1	5.000'−1		
	3	5.162'−1	5.162'−1	0.2581	
	4	2.681'−1	1.219'−1	0.5362	0.1219
	5	1.366'−1	3.271'−2	0.5291	0.1267
	6	6.978'−2	5.618'−3	0.5042	0.1786
	7	2.053'−2	−3.363'−4	0.5607	−0.1634
	8	4.547'−3	−3.484'−6	0.4772	−0.1556
	9	6.154'−4	1.325'−8	0.4296	−0.2144
	10	3.631'−5	−1.728'−12	0.3890	−0.2625
	11	9.956'−7	−3.844'−18	0.3558	−0.2477
	12	7.666'−9	−2.008'−26	0.3430	−0.2518
	13	1.215'−11		0.3360	
	14	2.548'−15		0.3339	
	15	3.104'−20		0.3334	
	16	1.032'−26		0.3333	
3	0	2.000	2.000		
	1	1.000	1.000		
	2	5.000'−1	5.000'−1		
	3	2.500'−1	2.500'−1		
	4	3.775'−1	3.775'−1	0.1887	
	5	1.814'−1	6.882'−2	0.3628	0.0688
	6	8.574'−2	1.567'−2	0.6860	0.1253
	7	4.214'−2	3.572'−3	0.4465	0.0757
	8	2.268'−2	7.222'−4	0.3313	0.1112
	9	5.580'−3	−3.949'−5	0.3588	−0.0970
	10	1.227'−3	−3.547'−7	0.3395	−0.0921
	11	2.347'−4	−2.893'−9	0.2455	−0.0716
	12	2.809'−5	8.630'−12	0.2219	−0.0847
	13	1.441'−6	−1.067'−15	0.2105	−0.1055
	14	5.518'−8	4.009'−21	0.1917	−0.0989
	15	1.164'−9		0.1766	
	16	7.021'−12		0.1735	
	17	1.354'−14		0.1703	
	18	1.077'−17		0.1677	
	19	1.365'−21		0.1670	

TABLE 9.1 (continued)

q	n	x_n	x'_n	r_n	r'_n
4	0	2.000	2.000		
	1	1.000	1.000		
	2	5.000'−1	5.000'−1		
	3	2.500'−1	2.500'−1		
	4	1.250'−1	1.250'−1		
	5	2.840'−1	2.840'−1	0.1420	
	6	1.258'−1	3.887'−2	0.2517	0.0389
	7	5.453'−2	7.030'−3	0.4362	0.0562
	8	2.492'−2	1.461'−3	0.7975	0.0935
	9	1.274'−2	4.448'−4	0.3588	0.0501
	10	7.507'−3	1.168'−4	0.2101	0.0846
	11	1.564'−3	−4.334'−6	0.2279	−0.0558
	12	3.227'−4	−2.390'−8	0.2374	−0.0598
	13	6.871'−5	−2.370'−10	0.2164	−0.0519
	14	1.360'−5	−2.500'−12	0.1423	−0.0329
	15	1.545'−6	9.027'−15	0.1316	−0.0401
	16	6.639'−8	−6.291'−19	0.1316	−0.0520
	17	2.814'−9	1.243'−24	0.1270	−0.0506
	18	1.067'−10		0.1142	
	19	2.207'−12		0.1050	
	20	1.073'−14		0.1046	
	21	1.944'−17		0.1040	
	22	3.069'−20		0.1022	
	23	2.367'−23		0.1005	

Table 9.1 was computed on an IBM 360/91 computer, with 14-digit truncated floating-point arithmetic to base 16. When computing the divided differences in equations (8.2) and (8.3), we took advantage of the fact that n-th divided differences of $1, x, x^2, \ldots, x^{n-1}$ vanish identically. Otherwise it is not possible to reduce $|x_n|$ or $|x'_n|$ to 10^{-20} without using higher precision arithmetic, because of the effect of rounding errors, except when $q = 1$.

For $q = 2$ our example is the same as that used by Jarratt (1967), and our results agree with his for $n \leq 9$. For $n = 10$ and 11 our results differ slightly, presumably because of rounding errors. The example given by Jarratt (1968) for $q = 3$ has also been verified.

Section 10
SUMMARY

The main results of this chapter for $q = 1$ (successive linear interpolation for finding a zero) and $q = 2$ (successive parabolic interpolation for finding a turning point) are summarized on p. 46.

THEOREM 3.1
$q = 1$: If $f \in C$ and $x_n \to \zeta$, then $f(\zeta) = 0$.
$q = 2$: If $f \in C^1$ and $x_n \dashrightarrow \zeta$, then $f'(\zeta) = 0$.

THEOREM 4.1
$q = 1$: If $f \in C^1, f'(\zeta) \neq 0$, and a good start, then superlinear convergence.
$q = 2$: If $f \in C^2, f''(\zeta) \neq 0$, and a good start, then superlinear convergence.

THEOREM 5.1
$q = 1$: If $f \in LC^1, f'(\zeta) \neq 0$, and a good start, then weak order at least
  $\beta_1 = 1.618\ldots.$
$q = 2$: If $f \in LC^2, f''(\zeta) \neq 0$, and a good start, then weak order at least
  $\beta_2 = 1.324\ldots.$

THEOREM 7.1
$q = 1$: If $f \in LC^2$, $f'(\zeta) \neq 0$, and a good start, then either strong order
  $\beta_1 = 1.618\ldots$ or weak order at least 2.
$q = 2$: If $f \in LC^3$, $f''(\zeta) \neq 0$, and a good start, then either strong order
  $\beta_2 = 1.324\ldots$ or weak order at least $[(3 + \sqrt{5})/2]^{1/3} = 1.378\ldots.$

THEOREM 8.1
$q = 1$: If $f \in LC^2, f'(\zeta) \neq 0$, and a good start, then the accelerated sequence
  converges with weak order at least $\beta'_1 = 1.839\ldots.$
$q = 2$: If $f \in LC^3, f''(\zeta) \neq 0$, and a good start, then the accelerated sequence
  converges with weak order at least $\beta'_2 = 1.465\ldots.$

4

AN ALGORITHM
WITH GUARANTEED
CONVERGENCE FOR FINDING
A ZERO OF A FUNCTION

Section 1
INTRODUCTION

Let f be a real-valued function, defined on the interval $[a, b]$, with $f(a)f(b) \leq 0$. f need not be continuous on $[a, b]$: for example, f might be a limited-precision approximation to some continuous function (see Forsythe (1969)). We want to find an approximation $\hat{\zeta}$ to a zero ζ of f, to within a given positive tolerance 2δ, by evaluating f at a small number of points. Of course, there may be no zero in $[a, b]$ if f is discontinuous, so we shall be satisfied if f takes both nonnegative and nonpositive values in $[\hat{\zeta} - 2\delta, \hat{\zeta} + 2\delta] \cap [a, b]$.

Clearly, such a $\hat{\zeta}$ may always be found by bisection in about $\log_2[(b-a)/\delta]$ steps, and this is the best that we can do for arbitrary f. In this chapter we describe an algorithm which is never much slower than bisection (see Section 3), but which has the advantage of superlinear convergence to a simple zero of a continuously differentiable function, if the effect of rounding errors is negligible. This means that, in practice, convergence is often much faster than for bisection (see Section 4). There is no contradiction here: bisection is the optimal algorithm (in a minimax sense) for the class of all functions which change sign on $[a, b]$, but it is not optimal for other classes of functions: e.g., C^1 functions with simple zeros, or convex functions. (See Gross and Johnson (1959), Bellman and Dreyfus (1962), and Chernousko (1970).)

47

Dekker's algorithm

The algorithm described here is similar to one, which we call Dekker's algorithm for short, variants of which have been given by van Wijngaarden, Zonneveld, and Dijkstra (1963); Wilkinson (1967); Peters and Wilkinson (1969); and Dekker (1969). We wish to emphasize that, although these variants of Dekker's algorithm have proved satisfactory in most practical cases, none of them guarantees convergence in less than about $(b - a)/\delta$ function evaluations. An example for which this bound is attained is given in Section 2. On the other hand, our algorithm must converge within about $\{\log_2[(b-a)/\delta]\}^2$ function evaluations (see Section 3). Typical values are $b - a = 1$ and $\delta = 10^{-12}$, giving 10^{12} and 1600 function evaluations respectively. Our point of view is that 1600 is a reasonable number, but 10^{12} is not, for a computer program which attempts to evaluate a function 10^{12} times is almost certain to run out of time.

On well-behaved functions, e.g., polynomials of moderate degree with well-separated zeros, both our algorithm and Dekker's are much faster than bisection. Our algorithm is at least as fast as Dekker's, and often slightly faster (see Section 4), so the only price to pay for the improvement in the guaranteed rate of convergence is a slight increase in the complexity of the algorithm.

Section 2
THE ALGORITHM

The algorithm is defined precisely by the ALGOL 60 procedure *zero* given in Section 6. Here we describe the algorithm, but the ALGOL procedure should be referred to for points of detail. For the motivation behind both our algorithm and Dekker's algorithm, see Dekker (1969) or Wilkinson (1967).

At a typical step we have three points a, b, and c such that $f(b)f(c) \leq 0$, $|f(b)| \leq |f(c)|$, and a may coincide with c. The points a, b, and c change during the algorithm, but there should be no confusion if we omit subscripts. b is the best approximation so far to ζ, a is the previous value of b, and ζ must lie between b and c. (Initially $a = c$.)

If $f(b) = 0$ then we are finished. The ALGOL procedure given by Dekker (1969) does not recognize this case, and can take a large number of small steps if f vanishes on an interval, which may happen because of underflow. This occurred with $f(x) = x^9$ on an IBM 360 computer.

If $f(b) \neq 0$, let $m = \frac{1}{2}(c - b)$. We prefer not to return with $\zeta = \frac{1}{2}(b + c)$ as soon as $|m| \leq 2\delta$, for if superlinear convergence has set in then b, the most recent approximation, is probably a much better approximation to

ζ than $\frac{1}{2}(b + c)$ is. Instead, we return with $\hat{\zeta} = b$ if $|m| \leq \delta$ (so the error is no more than δ if, as is often true, f is nearly linear between b and c), and otherwise interpolate or extrapolate f linearly between a and b, giving a new point i. (See later for inverse quadratic interpolation.) To avoid the possibility of overflow or division by zero, we find numbers p and q such that $i = b + p/q$, and the division is not performed if $2|p| \geq 3|mq|$, for then i is not needed anyway. The reason why the simpler criterion $|p| \geq |mq|$ is not used is explained later. Since $0 < |f(b)| \leq |f(a)|$ (see later), we can safely compute $s = f(b)/f(a)$, $p = \pm(a - b)s$, and $q = \mp(1 - s)$.

$$\text{Define} \quad b'' = \begin{cases} i \text{ if } i \text{ lies between } b \text{ and } b + m \text{ (``interpolation''),} \\ b + m \text{ otherwise (``bisection''),} \end{cases}$$

$$\text{and} \quad b' = \begin{cases} b'' \text{ if } |b - b''| > \delta, \\ b + \delta \, \text{sign}(m) \text{ otherwise (a ``step of } \delta\text{'').} \end{cases}$$

Dekker's algorithm takes b' as the next point at which f is evaluated, forms a new set $\{a, b, c\}$ from the old set $\{b, c, b'\}$, and continues. Unfortunately, it is easy to construct a function f for which steps of δ are taken every time, so about $(b - a)/\delta$ function evaluations are required for convergence. For example, let

$$f(x) = \begin{cases} 2^{x/\delta} \text{ for } a + \delta \leq x \leq b, \\ -\left(\dfrac{b - a - \delta}{\delta}\right)2^{b/\delta} \text{ for } x = a, \\ \text{arbitrary for } a < x < a + \delta. \end{cases} \qquad (2.1)$$

The first linear interpolation gives the point $b - \delta$, the next (an extrapolation) gives $b - 2\delta$, the next $b - 3\delta$, and so on.

Even if steps of δ are avoided, the asymptotic rate of convergence of successive linear interpolation may be very slow if f has a zero of sufficiently high multiplicity. (Note that none of the theorems of Chapter 3, apart from Theorem 3.3.1, apply for a multiple zero.) Suppose that $f \in C^n[a, b]$, $n > 1$, $\zeta \in (a, b)$, $f(\zeta) = f'(\zeta) = \ldots = f^{(n-1)}(\zeta) = 0$, and $f^{(n)}(\zeta) \neq 0$ (i.e., ζ is a root of multiplicity $n > 1$). If $\epsilon > 0$, $(x_1 - \zeta)/(x_0 - \zeta) \in (\epsilon, 1 - \epsilon)$, and x_0 is sufficiently close to ζ, then successive linear interpolation gives a sequence (x_n) which converges linearly to ζ. In fact, equation (3.2.1) holds with $\rho = 1$ and $K = \beta_{n-1}^{-1}$, where the constants $\beta_q \simeq 2^{2/(2q+1)}$ are defined in Definition 3.5.1. The proof is simple: if

$$y_m = \frac{x_{m+1} - \zeta}{x_m - \zeta} \qquad (2.2)$$

is the ratio of successive errors, then a Taylor series expansion of f about ζ gives

$$y_{m+1} = \left(\frac{1 - y_m^{n-1}}{1 - y_m^n}\right)(1 + o(1)) \qquad (2.3)$$

as $x_m \to \zeta$, provided y_m remains bounded away from 1. Now the iteration

$$z_{m+1} = g(z_m), \tag{2.4}$$

where

$$g(z) = \frac{1 - z^{n-1}}{1 - z^n}, \tag{2.5}$$

has fixed point $z = \beta_{n-1}^{-1}$, and

$$|g'(z)| < 1 \tag{2.6}$$

for $z \in (0, 1)$. Thus, the result follows from Theorem 22.1 of Ostrowski (1966).

An example for which convergence is sublinear (Definition 3.2.2) is

$$f(x) = \begin{cases} 0 & \text{if } x = 0, \\ x \cdot \exp(-x^{-2}) & \text{if } x \neq 0, \end{cases} \tag{2.7}$$

on an interval containing the origin. This is an extreme case, for f and all its derivatives vanish at the origin. (As a function of a complex variable, f has an essential singularity at the origin.) If

$$0 < x_1 < x_0 < \sqrt{2}, \tag{2.8}$$

then (x_n) is a positive, monotonic decreasing sequence, and, by Theorem 3.3.1, its limit must be 0. Thus, successive linear interpolation does converge, but very slowly.

Some of the examples above are rather artificial, and unless an extended exponent range is used (see later) we may be saved by underflow, i.e., the algorithm may terminate with a "zero" as soon as underflow occurs. Even so, it is clear that convergence may occasionally be very slow if Dekker's algorithm is used.

Our main modification of Dekker's algorithm ensures that a bisection is done at least once in every $2 \log_2(|b - c|/\delta)$ consecutive steps. The modification is this: let e be the value of p/q at the step before the last one. If $|e| < \delta$ or $|p/q| \geq \frac{1}{2}|e|$ then we do a bisection, otherwise we do either a bisection or an interpolation just as in Dekker's algorithm. Thus, $|e|$ decreases by at least a factor of two on every second step, and when $|e| < \delta$ a bisection must be done. (After a bisection we take $e = m$ for the next step.) This is why our algorithm, unlike Dekker's, is never much slower than bisection.

A simpler idea is to take e as the value of p/q at the last step, but practical tests show that this slows down convergence for well-behaved functions by causing unnecessary bisections. With the better choice of e, our experience has been that convergence is always at least as fast as for Dekker's algorithm (see Section 4).

Inverse quadratic interpolation

If the three current points a, b, and c are distinct, we can find the point i by inverse quadratic interpolation, i.e., fitting x as a quadratic in y, instead of by linear interpolation using just a and b. Experiments show that, for well-behaved functions, this device saves about 0.5 function evaluations per zero on the average (see Section 4). Inverse interpolation is used because with direct quadratic interpolation we have to solve a quadratic equation for i, and there is the problem of which root should be accepted. Cox (1970) gives another way of avoiding this problem: fit y as a function of the form $p(x)/q(x)$, where p and q are polynomials and p is linear. A third possibility is to use the acceleration technique described in Section 3.8. (See also Ostrowski (1966), Chapter 11.)

Care must be taken to avoid overflow or division by zero when computing the new point i. Since b is the most recent approximation to the root ζ, and a is the previous value of b, we do a bisection if $|f(b)| \geq |f(a)|$. Otherwise we have $|f(b)| < |f(a)| \leq |f(c)|$, so a safe way to find i is to compute $r_1 = f(a)/f(c)$, $r_2 = f(b)/f(c)$, $r_3 = f(b)/f(a)$, $p = \pm r_3[(c - b)r_1(r_1 - r_2) - (b - a)(r_2 - 1)]$, and $q = \mp(r_1 - 1)(r_2 - 1)(r_3 - 1)$. Then $i = b + p/q$, but as before we do not perform the division unless it is safe to do so. (If a bisection is to be done then i is not needed anyway.) When inverse quadratic interpolation is used the interpolating parabola cannot be a good approximation to f unless it is single-valued between $(b, f(b))$ and $(c, f(c))$. Thus, it is natural to accept the point i if it lies between b and c, and up to three-quarters of the way from b to c: consider the limiting case where the interpolating parabola has a vertical tangent at c and $f(b) = -f(c)$. Hence, we reject i if $2|p| \geq 3|mq|$.

The tolerance

As in Peters and Wilkinson (1969), the tolerance (2δ) is a combination of a relative tolerance (4ϵ) and an absolute tolerance ($2t$). At each step we take

$$\delta = 2\epsilon|b| + t, \tag{2.9}$$

where b is the current best approximation to ζ, $\epsilon = macheps$ is the *relative machine precision* ($\beta^{1-\tau}$ for τ-digit truncated floating-point arithmetic with base β, and half this for rounded arithmetic), and t is a positive absolute tolerance. Since δ depends on b, which could lie anywhere in the given interval, we should replace δ by its (positive) minimum over the interval in the upper bound for the number of function evaluations required. In the ALGOL procedures the variable *tol* is used for δ.

The effect of rounding errors

The ALGOL procedures given in Section 6 have been written so that rounding errors in the computation of i, m etc. cannot prevent convergence with the above choice of δ. The number 2ϵ in (2.9) may be increased if a higher relative error is acceptable, but it should not be decreased, for then rounding errors might prevent convergence.

The bound for $|\hat{\zeta} - \zeta|$ has to be increased slightly if we take rounding errors into account. Suppose that, for floating-point numbers x and y, the computed arithmetic operations satisfy

$$fl(x \times y) = xy(1 + \epsilon_1) \qquad (2.10)$$

and

$$fl(x \pm y) = x(1 + \epsilon_2) \pm y(1 + \epsilon_3), \qquad (2.11)$$

where $|\epsilon_i| \leq \epsilon$ for $i = 1, 2, 3$ (see Wilkinson (1963)). Also suppose that $fl(|x|) = |x|$ exactly, for any floating-point number x. The algorithm computes approximations

$$\tilde{m} = fl(0.5 \times (c - b)) \qquad (2.12)$$

and

$$\tilde{tol} = fl(2 \times \epsilon \times |b| + t), \qquad (2.13)$$

terminating only when

$$|\tilde{m}| \leq \tilde{tol} \qquad (2.14)$$

(unless $f(b) = 0$, when $\hat{\zeta} = \zeta = b$). Our assumptions (2.10) and (2.11) give

$$|\tilde{m}| \geq \tfrac{1}{2}(|c - b| - \epsilon(|b| + |c|))(1 - \epsilon) \qquad (2.15)$$

and

$$\tilde{tol} \leq (2\epsilon|b| + t)(1 + \epsilon)^3, \qquad (2.16)$$

so (2.14) implies that

$$|c - b| \leq \left(\frac{2}{1 - \epsilon}\right)(2\epsilon|b| + t)(1 + \epsilon)^3 + \epsilon(|b| + |c|). \qquad (2.17)$$

Since $|\hat{\zeta} - \zeta| \leq |c - b|$ and $b = \hat{\zeta}$, this gives

$$|\hat{\zeta} - \zeta| \leq 6\epsilon|\zeta| + 2t, \qquad (2.18)$$

neglecting terms of order ϵt and $\epsilon^2|\zeta|$. Usually the error is less than half this bound (see above).

Of course, it is the user's responsibility to consider the effect of rounding errors in the computation of f. The ALGOL procedures only guarantee to find a zero ζ of the *computed* function f to an accuracy given by (2.18), and ζ may be nowhere near a root of the mathematically defined function that the user is really interested in!

Extended exponent range

In some applications the range of f may be larger than is allowed for standard floating-point numbers. For example, $f(x)$ might be $\det(A - xI)$, where A is a matrix whose eigenvalues are to be found. In Section 6 we give an ALGOL procedure (*zero2*) which accepts $f(x)$ represented as a pair

$(y(x), z(x))$, where $f(x) = y(x) \cdot 2^{z(x)}$ (y real, z integer). Thus, *zero2* will accept functions in the same representation as is assumed by Peters and Wilkinson (1969), although *zero2* does not require that $1/16 \leq |y(x)| < 1$ (unless $y(x) = 0$), and could be simplified slightly if this assumption were made.

Section 3
CONVERGENCE PROPERTIES

If the initial interval is $[a, b]$, assume that

$$b - a > \delta_m, \tag{3.1}$$

and let

$$k = \lceil \log_2((b - a)/\delta_m) \rceil, \tag{3.2}$$

where δ_m is the minimum over $[a, b]$ of the tolerance

$$\delta(x) = 2macheps|x| + t \tag{3.3}$$

(see Section 2), and $\lceil x \rceil$ means the least integer $y \geq x$. By assumption (3.1), $k > 0$. (Procedure *zero* takes only two function evaluations if $k \leq 0$.)

First consider a bisection process terminating when the interval known to contain a zero has length $\leq 2\delta_m$ (so the endpoint minimizing $|f|$ is probably within δ_m of the zero, and certainly within $2\delta_m$). It is easy to see that this process terminates after exactly $k + 1$ function evaluations unless, by good fortune, f happens to vanish at one of the points of evaluation.

Now consider procedure *zero* or *zero2*. If $k = 1$ then the procedure terminates after two function evaluations, one at each end-point of the initial interval. If $k = 2$ then there are two initial evaluations, and after no more than four more evaluations a bisection must be done, for the reason described in Section 2. After this bisection, which requires one more function evaluation, the procedure must terminate. Thus, at most $2 + 5 = 7$ evaluations are required. Similarly, for $k \geq 1$, the maximum number of function evaluations required is

$$2 + (5 + 7 + 9 + \ldots + (2k + 1)) = (k + 1)^2 - 2. \tag{3.4}$$

Since Dekker's algorithm may take up to 2^k function evaluations (see Section 2), this justifies the remarks made in Section 1. Also, although the upper bound (3.4) is attainable, it is clear that it is unlikely to be attained except for very contrived examples, and in practical tests our algorithm has never taken more than $3(k + 1)$ function evaluations (see Section 4). This justifies the claim that our algorithm is never much slower than bisection.

Superlinear convergence

Ignoring the effect of rounding errors and the tolerance δ, we see, as in Dekker (1969), that the algorithm will eventually stop doing bisections when it is approaching a simple zero ζ of a C^1 function. Thus, temporarily ignoring the improvement described in Section 2, the theorems of Chapter 3 are applicable (with $q = 1$). In particular, convergence is superlinear, in the sense that $\lim_{n \to \infty} |x_n - \zeta|^{1/n} = 0$, and equation (3.4.22) holds (see Theorem 3.4.1). If f' is Lipschitz continuous near ζ, then the weak order of convergence is at least $\frac{1}{2}(1 + \sqrt{5}) = 1.618\ldots$ (Theorem 3.5.1).

If f' is Lipschitz continuous near the simple zero ζ then, even with the inverse parabolic interpolation modification described in Section 2, the weak order of convergence is still at least $\frac{1}{2}(1 + \sqrt{5})$. The idea of the proof is that, by Lemma 2.5.1, the curvature at ζ of the approximating parabolas is bounded, so the inequality (3.5.13) still holds for some M (no longer the Lipschitz constant) and sufficiently small δ_n.

Thus, our procedure always converges in a reasonable number of steps and, under the conditions mentioned above, convergence is superlinear with order at least $1.618\ldots$. It is well known that, since $(1.618\ldots)^2 = 2.618\ldots > 2$, this compares favorably with Newton's method if an evaluation of f' is as expensive as an evaluation of f. In practice, convergence for well-behaved functions is fast, and the stopping criterion is usually satisfied in a few steps once superlinear convergence sets in.

Section 4
PRACTICAL TESTS

The ALGOL procedures *zero* (for standard floating-point numbers) and *zero2* (for floating-point with an extended exponent range) have been tested using ALGOL W (Wirth and Hoare (1966), Bauer, Becker, and Graham (1968)) on IBM 360/67 and 360/91 computers with machine precision 16^{-13}. The number of function evaluations for convergence has never been greater than three times the number required for bisection, even for the functions given by (2.1) and (2.7), and for these functions Dekker's algorithm takes more than 10^6 function evaluations. *Zero2* has been tested extensively with eigenvalue routines, and in this application it usually takes the same or one less function evaluation per eigenvalue than Dekker's algorithm, and considerably less than bisection.

In Table 4.1, we give the number of function evaluations required for convergence with procedure *zero2* and functions x^9, x^{19}, $f_1(x)$, and $f_2(x)$, where

$$f_1(x) = \begin{cases} 0 & \text{if } |x| < 3.8 \times 10^{-4} \\ fl(x \exp(-x^{-2})) & \text{otherwise,} \end{cases} \tag{4.1}$$

and

$$f_2(x) = \begin{cases} fl(\exp(x)) & \text{if } x > -10^6, \\ fl(\exp(-10^6) - (x + 10^6)^2) & \text{otherwise.} \end{cases} \tag{4.2}$$

The parameters a, b, and t of procedure $zero2$ are given in the table. In all cases $macheps = 16^{-13}$.

TABLE 4.1 **The number of function evaluations for convergence with procedure**
$zero2$

$f(x)$	a	b	t	$\hat{\zeta} - \zeta$	Function Evals.
x^9	-1.0	$+1.1$	$1'-9$	$4.99'-10$	81
x^9	-1.0	$+4.0$	$1'-20$	$4.92'-21$	189
x^{19}	-1.0	$+4.0$	$1'-20$	$4.81'-21$	195
$f_1(x)$	-1.0	$+4.0$	$1'-20$	0^*	33
$f_2(x)$	-1001200	0	$1'-20$	$1'-9$	79

$^*\hat{\zeta} = 2.17'-4$ and $f_1(\hat{\zeta}) = 0$.

In Table 4.2, we compare the procedure given by Dekker (1969) with procedure $zero$ (procedure $zero2$ gives identical results as no underflow or overflow occurs) for a typical application: finding the eigenvalues of a symmetric band matrix by repeated determinant evaluation. Let A be the n by n 5-diagonal matrix defined by

$$a_{ij} = \begin{cases} p - r & \text{if } i = j = 1 \text{ or } i = j = n, \\ p & \text{if } 1 < i = j < n, \\ 2q & \text{if } |i - j| = 1, \\ r & \text{if } |i - j| = 2, \\ 0 & \text{if } |i - j| > 2. \end{cases} \tag{4.3}$$

For $n > 2$, A has eigenvalues

$$\lambda_k = p - 4q \cdot \cos\left(\frac{k\pi}{n+1}\right) + 2r \cdot \cos\left(\frac{2k\pi}{n+1}\right) \tag{4.4}$$

for $k = 1, 2, \ldots, n$ (Ehrlich (1970)). Table 4.2 gives the eigenvalues λ_k, the number n_D of function evaluations per eigenvalue for Dekker's procedure, and the number n_Z of function evaluations for procedure $zero$. For each eigenvalue, the tolerances for Dekker's procedure and for procedure $zero$ were the same. (The tolerance was adjusted by the eigenvalue program to ensure that the computed eigenvalues had a relative error of less than 5×10^{-14}.) Tests were run for several values of n, p, q, and r: the table gives a typical set of results for $n = 15$, $p = 7$, $q = 7/4$, and $r = 1/2$. To obtain the same accuracy with bisection, at least 40 function evaluations per eigenvalue would be required, so both our procedure and Dekker's are at least four times as fast as bisection for this application.

TABLE 4.2 Comparison of Dekker's procedure with procedure *zero**

k	λ_k	n_D	n_Z
1	1.05838256968867	10	10
2	1.23995005360754	10	9
3	1.56239614624727	10	10
4	2.05025253169417	10	10
5	2.72832493649769	11	10
6	3.61410919225782	11	10
7	4.71048821337581	10	10
8	6.00000000000000	9	9
9	7.44175272160161	10	9
10	8.97167724536908	10	10
11	10.5063081987721	10	10
12	11.9497474683058	10	9
13	13.2029707184829	10	9
14	14.1742635087655	10	9
15	14.7893764953339	9	8

*For a definition of λ_k, n_D, and n_Z, see above. The λ_k have a relative error of less than 5'–14.

Some more experimental results are given in Chapter 5. For an illustration of the superlinear convergence, see the examples given in Section 3.9.

Section 5
CONCLUSION

Our algorithm appears to be at least as fast as Dekker's on well-behaved functions and, unlike Dekker's, it is guaranteed to converge in a reasonable number of steps for any function. The ALGOL procedures *zero* and *zero2* given in Section 6 have been written to avoid problems with rounding errors or overflow, and floating-point underflow is not harmful as long as the result is set to zero.

Before giving the ALGOL procedures *zero* and *zero2*, we briefly discuss some possible extensions.

Cox's algorithm

Cox (1970) gives an algorithm which combines bisection with interpolation, using both f and f'. This algorithm may fail to converge in a reasonable number of steps in the same way as Dekker's. A simple modification, exactly like the one that we have given in Section 2 for Dekker's algorithm, will remedy this defect without slowing the rate of convergence for well-behaved functions.

Parallel algorithms

In this chapter we have considered only serial algorithms. It is well known (see, for example, Traub (1964)) that all serial methods which use only function evaluations and Lagrange interpolating polynomials have weak order less than 2, unless certain relations hold between the derivatives of f at ζ. Winograd and Wolfe (1971) have shown that no serial method, using only function evaluations, can have order greater than 2 for all analytic functions with simple zeros. Thus, nothing much can be gained by going beyond linear or quadratic interpolation. However, Miranker (1969) has shown that, if a parallel computer is available, a class of algorithms using Lagrange interpolating polynomials gives superlinear convergence with weak order greater than 2 under certain conditions. Also, it is possible to generalize the bisection process to "$(r + 1)$-section" with advantage if a parallel computer with r independent processors is available. See, for example, Wilde (1964). There does not appear to be any fundamental difficulty in combining generalized bisection with one of Miranker's parallel algorithms so that convergence in a reasonable number of steps is guaranteed for any function, and superlinear convergence with order greater than 2 is likely for well-behaved functions.

Searching an ordered file

A problem which is commonly solved by a binary search (i.e., bisection) method is that of locating an element in a large ordered file. The problem may be formalized in the following way. Let S be a totally ordered set, and $\varphi \colon S \to R$ an order-preserving mapping from S into the real numbers. Suppose that $T = \{t_0, t_1, \ldots, t_n\}$ is a finite subset of S, with $t_0 < t_1 < \ldots < t_n$. Given $c \in [\varphi(t_0), \varphi(t_n)]$, we may define a monotonic function f on $[0, n]$ by

$$f(x) = \varphi(t_i) - c, \qquad (5.1)$$

where $x \in [0, n]$ and $i = \lceil x - \tfrac{1}{2} \rceil$. Thus, finding an index i such that $\varphi(t_i) = c$ is equivalent to finding a zero of f in $[0, n]$, and our zero-finding algorithm could be used instead of the usual bisection algorithm. It might be worthwhile to modify our algorithm slightly to take the discrete nature of the problem into account.

Section 6

ALGOL 60 PROCEDURES

The ALGOL procedures *zero* (for standard floating-point numbers) and *zero2* (for floating-point with an extended exponent range) are given below. For a description of the idea of the algorithm, see Section 2. Some test cases and numerical results are described in Section 4. A FORTRAN translation of procedure *zero* is given in the Appendix.

real procedure *zero* $(a, b, macheps, t, f)$;
value $a, b, macheps, t$; **real** $a, b, macheps, t$; **real procedure** f;
  begin comment:
        Procedure *zero* returns a zero x of the function f in the given interval $[a, b]$, to within a tolerance $6macheps|x| + 2t$, where *macheps* is the relative machine precision and t is a positive tolerance. The procedure assumes that $f(a)$ and $f(b)$ have different signs;
  real $c, d, e, fa, fb\ fc, tol, m, p, q, r, s$;
  $fa: = f(a); fb: = f(b)$;
  int: $c: = a; fc: = fa; d: = e: = b - a$;
  ext: **if** abs$(fc) <$ abs(fb) **then**
        begin $a: = b; b: = c; c: = a$;
        $fa: = fb; fb: = fc; fc: = fa$
        end;
  $tol: = 2 \times macheps \times$ abs$(b) + t; m: = 0.5 \times (c - b)$;
  if abs$(m) > tol \wedge fb \neq 0$ **then**
        begin comment: See if a bisection is forced;
        if abs$(e) < tol \vee$ abs$(fa) \leq$ abs(fb) **then** $d: = e: = m$ **else**
            begin $s: = fb/fa$; **if** $a = c$ **then**
                begin comment: Linear interpolation;
                $p: = 2 \times m \times s; q: = 1 - s$
                end
            else
                begin comment: Inverse quadratic interpolation;
                $q: = fa/fc; r: = fb/fc$;
                $p: = s \times (2 \times m \times q \times (q - r) - (b - a) \times (r - 1))$;
                $q: = (q - 1) \times (r - 1) \times (s - 1)$
                end;
            if $p > 0$ **then** $q: = -q$ **else** $p: = -p$;
            $s: = e; e: = d$;
            if $2 \times p < 3 \times m \times q -$ abs$(tol \times q) \wedge p <$ abs$(0.5 \times s \times q)$
            then $d: = p/q$ **else** $d: = e: = m$
            end;
        $a: = b; fa: = fb$;
        $b: = b + ($**if** abs$(d) > tol$ **then** d **else if** $m > 0$ **then** tol **else** $-tol)$;

$fb: = f(b);$

go to if $fb > 0 \equiv fc > 0$ **then** int **else** ext

end;

$zero: = b$

end $zero;$

real procedure $zero2$ $(a, b, macheps, t, f);$

value $a, b, macheps, t;$ **real** $a, b, macheps, t;$ **procedure** $f;$

begin comment:

Procedure $zero2$ finds a zero of the function $\bar{f}$ in the same way as procedure $zero$, except that the procedure $f(x, y, z)$ returns y(real) and z (integer) so that $\bar{f}(x) = y \cdot 2^z$. Thus underflow and overflow can be avoided with a very large function range;

real procedure $pwr2$ $(x, n);$ **value** $x, n;$ **real** $x;$ **integer** $n;$

comment: This procedure is machine-dependent. It computes $x \cdot 2^n$ for $n \leq 0$, avoiding underflow in the intermediate results;

$pwr2: =$ **if** $n > -200$ **then** $x \times 2 \uparrow n$ **else**

if $n > -400$ **then** $(x \times 2 \uparrow (-200)) \times 2 \uparrow (n + 200)$ **else**

if $n > -600$ **then** $((x \times 2 \uparrow (-200)) \times 2 \uparrow (-200)) \times 2 \uparrow (n + 400)$

else $0;$

integer $ea, eb, ec;$

real $c, d, e, fa, fb, fc, tol, m, p, q, r, s;$

$f(a, fa, ea); f(b, fb, eb);$

int: $c: = a; fc: = fa; ec: = ea; d: = e: = b - a;$

ext: **if** $(ec \leq eb \wedge pwr2(abs(fc), ec - eb) < abs(fb))$

$\vee$ $(ec > eb \wedge pwr2(abs(fb), eb - ec) \geq abs(fc))$ **then**

begin $a: = b; fa: = fb; ea: = eb;$

$b: = c; fb: = fc; eb: = ec;$

$c: = a; fc: = fa; ec: = ea$

end;

$tol: = 2 \times macheps \times abs(b) + t; m: = 0.5 \times (c - b);$

if $abs(m) > tol \wedge fb \neq 0$ **then**

begin if $abs(e) < tol \vee$

$(ea \leq eb \wedge pwr2(abs(fa), ea - eb) \leq abs(fb)) \vee$

$(ea > eb \wedge pwr2(abs(fb), eb - ea) \geq abs(fa))$ **then**

$d: = e: = m$ **else**

begin $s: = pwr2(fb, eb - ea)/fa;$ **if** $a = c$ **then**

begin $p: = 2 \times m \times s; q: = 1 - s$ **end**

else

begin $q: = pwr2(fa, ea - ec)/fc;$

$r: = pwr2(fb, eb - ec)/fc;$

$p: = s \times (2 \times m \times q \times (q - r) - (b - a) \times (r - 1));$

$q: = (q - 1) \times (r - 1) \times (s - 1)$

end;

```
        if p > 0 then q: = −q else p: = −p; s: = e; e: = d;
        if 2 × p < 3 × m × q − abs(tol × q) ∧
        p < abs(0.5 × s × q) then
        d: = p/q else d: = e: = m
        end;
    a: = b; fa: = fb; ea: = eb;
    b: = b + (if abs(d) > tol then d else if m > 0 then tol else − tol);
    f(b, fb, eb);
    go to if fb > 0 ≡ fc > 0 then int else ext
    end;
zero2: = b
end zero2;
```

5

AN ALGORITHM WITH GUARANTEED CONVERGENCE FOR FINDING A MINIMUM OF A FUNCTION OF ONE VARIABLE

Section 1
INTRODUCTION

A common computational problem is finding an approximation to the minimum or maximum of a real-valued function f in some interval $[a, b]$. This problem may arise directly or indirectly. For example, many methods for minimizing functions $g(\mathbf{x})$ of several variables need to minimize functions of one variable of the form

$$\gamma(\lambda) = g(\mathbf{x}_0 + \lambda \mathbf{s}), \qquad (1.1)$$

where $\mathbf{x}_0$ and $\mathbf{s}$ are fixed (a "one-dimensional search" from $\mathbf{x}_0$ in the direction $\mathbf{s}$). In this chapter we give an algorithm which finds an approximate local minimum of f by evaluating f at a small number of points. There is a clear analogy between this algorithm and the algorithm for zero-finding described in Chapter 4 (see Section 4). Unless f is unimodal (Section 3), the local minimum may not be the global minimum of f in $[a, b]$, and the problem of finding global minima is left until Chapter 6.

The algorithm described in this chapter could be used to solve the problem (1.1), but it would be more economical to use special algorithms which make use of any extra information which is available (e.g., estimates of the second derivative of γ), and which do not attempt to find the minimum very accurately. This is discussed in Chapter 7. Thus, a more likely practical use for our algorithm is to find accurate minima of naturally arising functions of one variable.

In Section 2 we consider the effect of rounding errors on any minimization algorithm based entirely on function evaluations. Unimodality is defined in Section 3, and we also define "δ-unimodality" in an attempt to explain why methods like golden section search work even for functions which are not quite unimodal (because of rounding errors in their computation, for example). In Sections 4 and 5 we describe a minimization algorithm analogous to the zero-finding algorithm of Chapter 4, and some numerical results are given in Section 6. Finally, some possible extensions are described in Section 7, and an ALGOL 60 procedure is given in Section 8.

Reduction to a zero-finding problem

If f is differentiable in $[a, b]$, a necessary condition for f to have a local minimum at an interior point $\mu \in (a, b)$ is

$$f'(\mu) = 0. \tag{1.2}$$

There is also the possibility that the minimum is at a or b: for example, this is true if f' does not change sign on $[a, b]$. If we are prepared to check for this possibility, one approach is to look for zeros of f'. If f' has different signs at a and b, then the algorithm of Chapter 4 may be used to approximate a point μ satisfying (1.2).

Since f' vanishes at any stationary point of f, it is possible that the point found is a maximum, or even an inflexion point, rather than a minimum. Thus, it is necessary to check whether the point found is a true minimum, and continue the search in some way if it is not.

If it is difficult or impossible to compute f' directly, we could approximate f' numerically (e.g., by finite differences), and search for a zero of f' as above. However, a method which does not need f' seems more natural, and could be preferred for the following reasons:

1. It may be difficult to approximate f' accurately because of rounding errors;
2. A method which does not need f' may be more efficient (see below); and
3. Whether f' can be computed directly or not, a method which avoids difficulty with maxima and inflexion points is clearly desirable.

Jarratt's method

Jarratt (1967) suggests a method, using successive parabolic interpolation, which is a special case of the iteration analyzed in Chapter 3. With arbitrary starting points Jarratt's method may diverge, or converge to a maximum or inflexion point, but this defect need not be fatal if the method is used in combination with a safe method such as golden section search, in the same way that we used a combination of successive linear interpolation and bisection for finding a zero. Theorem 3.5.1 shows that, if f has a Lipschitz continuous second derivative which is positive at an interior minimum μ, then Jarratt's method gives superlinear convergence to μ with weak order at least $\beta_2 = 1.3247 \ldots$ (see Definitions 3.2.1 and 3.5.1), provided the initial approximation is good and rounding errors are negligible.

Let us compare Jarratt's method with one of the alternatives: estimating f' by finite differences, and then using successive linear interpolation to find a zero of f'. (This process may also diverge, or converge to a maximum.) Suppose that $f''(\mu) > 0$ and $f^{(3)}(\mu) \neq 0$, to avoid exceptional cases (see Sections 3.6, 3.7, and 4.2). Since at least two function evaluations are needed to estimate f' at any point, and $\sqrt{1.618} \ldots = 1.272 \ldots < 1.324 \ldots$, Jarratt's method has a slightly higher order of convergence. The comparison is similar to that between Newton's method and successive linear interpolation: see Section 4.3 and Ostrowski (1966).

Section 2
FUNDAMENTAL LIMITATIONS BECAUSE
OF ROUNDING ERRORS

Suppose that $f \in LC^2[a, b; M]$ has a minimum at $\mu \in (a, b)$. Since $f'(\mu) = 0$, Lemma 2.3.1 gives, for $x \in [a, b]$,

$$f(x) = f_0 + \frac{1}{2}f_0''(x - \mu)^2 + \frac{m_x}{6}(x - \mu)^3, \tag{2.1}$$

where $|m_x| \leq M$, $f_0 = f(\mu)$, and $f_0'' = f''(\mu)$. Because of rounding errors, the best that can be expected if single-precision floating-point numbers are used is that the computed value $fl(f(x))$ of $f(x)$ satisfies the (nearly attainable) bound

$$fl(f(x)) = f(x)(1 + \epsilon_x), \tag{2.2}$$

where

$$|\epsilon_x| \leq \epsilon, \tag{2.3}$$

and ϵ is the relative machine precision (see Section 4.2). The error bound is unlikely to be as good as this unless f is a very simple function, or is evaluated using double-precision and then rounded or truncated to single-precision.

Let δ be the largest number such that, according to equations (2.2) and (2.3), it is possible that

$$fl(f(\mu + \delta)) \leq f_0. \tag{2.4}$$

It is unreasonable to expect any minimization procedure, based on single-precision evaluations of f, to return an approximation $\hat{\mu}$ to μ with a guaranteed upper bound for $|\hat{\mu} - \mu|$ less than δ. This is so regardless of whether the computed values of f are used directly, as in Jarratt's method, or indirectly, as in the other method suggested in Section 1. The reason is simply that the minimum of the computed function $fl(f(x))$ may lie up to a distance δ from the minimum μ of $f(x)$: see Diagram 2.1.

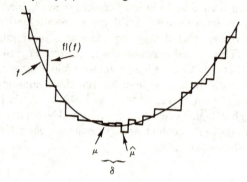

DIAGRAM 2.1   The effect of rounding errors

If $f_0'' > 0$, equations (2.1) to (2.4) give

$$\delta \geq \sqrt{\frac{2|f_0|\epsilon}{f_0''}}\Big(1 - \epsilon - \frac{M\delta}{6f_0''}\Big). \tag{2.5}$$

Thus, if $\mu \neq 0$ and the term $M\delta/(6f_0'')$ is negligible, an upper bound for the relative error $|(\hat{\mu} - \mu)/\mu|$ could hardly be less than $[2|f_0|\epsilon/(\mu^2 f_0'')]^{1/2}$, and full single-precision accuracy in $\hat{\mu}$ is unlikely unless $|f_0|/(\mu^2 f_0'')$ is of order ϵ or less, although $fl(f(\hat{\mu}))$ may agree with $f(\mu)$ to full single-precision accuracy. (See also Pike, Hill, and James (1967).)

If f' has a simple analytic representation, then it may be easy to compute f' accurately. For example, perhaps

$$fl(f'(x)) = f'(x(1 + \epsilon_x'))(1 + \epsilon_x''), \tag{2.6}$$

where $|\epsilon_x'| \leq \epsilon$ and $|\epsilon_x''| \leq \epsilon$, so we can expect to find a zero of f' with a relative error bounded by ϵ (see Lancaster (1966) and Ostrowski (1967b)). If (2.6) holds it might be worthwhile to use the algorithm described in Chapter 4 to search for a zero of f', or at least use it to refine the approximation $\hat{\mu}$ given by a procedure using only evaluations of f. However, this is not so if f' has to be approximated by differences, for then (2.6) cannot be expected to hold.

Even if $f(x)$ is a unimodal function, the computed approximation $fl(f(x))$ will not be unimodal: $fl(f(x))$ must be constant over small intervals of real numbers x which have the same floating-point approximation $fl(x)$. In the next section we define "δ-unimodality" to circumvent this difficulty.

From now on, we consider the problem of approximating the minimum of the *computed* function, or, equivalently, we ignore rounding errors in the computation of *f*. The user should bear in mind that the minimum of the computed function may differ from the minimum that he is really interested in by as much as δ (see equation (2.5) above). There is no point in wasting function evaluations by finding the minimum of the computed function to excessive accuracy, and our procedure *localmin* (Section 8) should not be called with the parameter *eps* much less than $[2\,|f_0|\,\epsilon/(\mu^2 f_0'')]^{1/2}$.

Section 3
UNIMODALITY AND δ-UNIMODALITY

There are several different definitions of a unimodal function in the literature. One source of confusion is that the definition depends on whether the function is supposed to have a unique minimum or a unique maximum (we consider minima). Kowalik and Osborne (1968) say that *f* is unimodal on [*a*, *b*] if *f* has only one stationary value on [*a*, *b*]. This definition has two disadvantages. First, it is meaningless unless *f* is differentiable on [*a*, *b*], but we would like to say that $|x|$ is unimodal on [−1, 1]. Second, functions which have inflexion points with a horizontal tangent are prohibited, but we would like to say that $f(x) = x^6 - 3x^4 + 3x^2$ is unimodal on [−2, 2] (here $f'(\pm 1) = f''(\pm 1) = 0$).

Wilde (1964) gives another definition: *f* is unimodal on [*a*, *b*] if, for all $x_1, x_2 \in [a, b]$,

$$x_1 < x_2 \supset (x_2 < x^* \supset f(x_1) > f(x_2)) \wedge (x_1 > x^* \supset f(x_1) < f(x_2)),$$

$$(3.1)$$

where x^* is a point at which *f* attains its least value in [*a*, *b*]. (We have reversed some of Wilde's inequalities as he considers maxima rather than minima.) Wilde's definition does not assume differentiability, or even continuity, but to verify that a function *f* satisfies (3.1) we need to know the point x^* (and such a point must exist). Hence, we prefer the following definition, which is nearly equivalent to Wilde's (see Lemma 3.1), but avoids any reference to the point x^*. The definition is not as complicated as it looks: it merely says that *f* cannot have a "hump" between any two points x_0 and x_2 in [*a*, *b*]. Two possible configurations of the points x_0, x_1, x_2, and x^* in (3.1) and (3.2) are illustrated in Diagram 3.1.

DEFINITION 3.1

f is *unimodal* on [*a*, *b*] if, for all x_0, x_1 and $x_2 \in [a, b]$,

$$x_0 < x_1 \wedge x_1 < x_2 \supset (f(x_0) \leq f(x_1) \supset f(x_1) < f(x_2)) \wedge$$
$$(f(x_1) \geq f(x_2) \supset f(x_0) > f(x_1)).$$

$$(3.2)$$

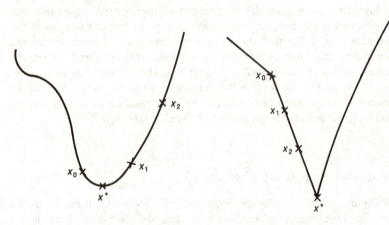

DIAGRAM 3.1 Unimodal functions

LEMMA 3.1

If a point x^* at which f attains its minimum in $[a, b]$ exists, then Wilde's definition of unimodality and Definition 3.1 are equivalent.

Proof

Suppose that f is unimodal according to Definition 3.1. If $x_1 < x_2$ and $x_2 < x^*$, take $x'_0 = x_1$, $x'_1 = x_2$, and $x'_2 = x^*$. Since f attains its least value at x^*,

$$f(x'_1) \geq f(x^*) = f(x'_2), \tag{3.3}$$

so equation (3.2) with primed variables gives

$$f(x'_0) > f(x'_1), \tag{3.4}$$

and thus

$$f(x_1) > f(x_2). \tag{3.5}$$

Similarly, if $x_1 < x_2$ and $x_1 > x^*$, equation (3.2) gives

$$f(x_1) < f(x_2). \tag{3.6}$$

Thus, from (3.5) and (3.6), equation (3.1) holds.

Conversely, suppose that (3.1) holds and $x_0 < x_1 < x_2$. If $f(x_0) \leq f(x_1)$ then there are three possibilities, depending on the position of x^*:

1. $x_1 > x^*$. Thus, by (3.1),

$$f(x_1) < f(x_2). \tag{3.7}$$

2. $x_1 = x^*$. Take $x'_1 = \frac{1}{2}(x_1 + x_2)$ and $x'_2 = x_2$.

Since $x^* < x'_1 < x'_2$, equation (3.1) with primed variables gives

$$f(x'_1) < f(x'_2), \tag{3.8}$$

so

$$f(x_1) = f(x^*) \leq f(x'_1) < f(x'_2) = f(x_2). \tag{3.9}$$

3. $x_1 < x^*$. Take $x_1' = x_0$ and $x_2' = x_1$. Since $x_1' < x_2' < x^*$, equation (3.1) gives $f(x_1') > f(x_2')$, contradicting the assumption that $f(x_0) \leq f(x_1)$. Hence case 3 is impossible and, by (3.7) and (3.9), we always have $f(x_1) < f(x_2)$.

Similarly, if $f(x_1) \geq f(x_2)$ then $f(x_0) > f(x_1)$, so equation (3.2) holds, and the proof is complete.

A simple corollary of Lemma 3.1 is that, if f is continuous, then Wilde's definition of unimodality and ours are equivalent. For arbitrary f the definitions are not equivalent. For example,

$$f(x) = \begin{cases} 1 - x & \text{if } x \leq 0, \\ x & \text{if } x > 0 \end{cases} \tag{3.10}$$

is unimodal on $[-1, 1]$ by our definition, but not by Wilde's, for x^* does not exist.

The following theorem gives a simple characterization of unimodality. There is no assumption that f is continuous. Since a strictly monotonic function (e.g., x^3) may have stationary points, the theorem shows that both our definition and Wilde's are essentially different from Kowalik and Osborne's, even if f is continuously differentiable. (Although this point is obvious, it is sometimes overlooked! See also Corollary 3.3.)

THEOREM 3.1

f is unimodal on $[a, b]$ (according to Definition 3.1) iff, for some (unique) $\mu \in [a, b]$, either f is strictly monotonic decreasing on $[a, \mu)$ and strictly monotonic increasing on $[\mu, b]$, or f is strictly monotonic decreasing on $[a, \mu]$ and strictly monotonic increasing on $(\mu, b]$.

The theorem is a special case of Theorem 3.2 below, so the proof is omitted. The following corollaries are immediate.

COROLLARY 3.1

If f is unimodal on $[a, b]$, then f attains its least value at most once on $[a, b]$. (If f attains its least value, then it must attain it at the point μ given by Theorem 3.1.)

COROLLARY 3.2

If f is unimodal and continuous on $[a, b]$, then f attains its least value exactly once on $[a, b]$.

COROLLARY 3.3

If $f \in C^1[a, b]$ then f is unimodal iff, for some $\mu \in [a, b]$, $f' < 0$ almost everywhere on $[a, \mu]$ and $f' > 0$ almost everywhere on $[\mu, b]$. (Note that f' may vanish at a finite number of points.)

Fibonacci and golden section search

If f is unimodal on $[a, b]$, then the minimum of f (or, if the minimum is not attained, the point μ given by Theorem 3.1) can be located to any desired accuracy by the well-known methods of Fibonacci search or golden section search. The reader is referred to Wilde (1964) for an excellent description of these methods. (See also Boothroyd (1965a, b), Johnson (1955), Krolak (1968), Newman (1965), Pike and Pixner (1967), and Witzgall (1969).) Care should be taken to ensure that the coordinates of the points at which f is evaluated are computed in a numerically stable way (see Overholt (1965)). Fibonacci and golden section search, as well as similar but less efficient methods, are based on the following result, which shows how an interval known to contain μ may be reduced in size.

COROLLARY 3.4

Suppose that f is unimodal on $[a, b]$, μ is the point given by Theorem 3.1, and $a \leq x_1 < x_2 \leq b$. If $f(x_1) \leq f(x_2)$ then $\mu \leq x_2$, and if $f(x_1) \geq f(x_2)$ then $\mu \geq x_1$.

Proof

If $x_2 < \mu$ then, by Theorem 3.1, $f(x_1) > f(x_2)$. Thus, if $f(x_1) \leq f(x_2)$ then $\mu \leq x_2$. The other half follows similarly.

If the reader is prepared to ignore the problem of computing unimodal functions using limited-precision arithmetic, he may skip the rest of this section.

δ-unimodality

We pointed out at the end of Section 2 that functions computed using limited-precision arithmetic are not unimodal. Thus, the theoretical basis for Fibonacci search and similar methods is irrelevant, and it is not clear that these methods will give even approximately correct results in the presence of rounding errors. To analyze this problem, we generalize the idea of unimodality to δ-unimodality. Intuitively, δ is a nonnegative number such that Fibonacci or golden section search will give correct results, even though f is not necessarily unimodal (unless $\delta = 0$), provided that the distance between points at which f is evaluated is always greater than δ. The results of Section 2 indicate how large δ is likely to be in practice. (Our aim differs from that of Richman (1968) in defining the ϵ-calculus, for he is interested in properties that hold as $\epsilon \to 0$.) For another approach to the problem of rounding errors, see Overholt (1967).

In the remainder of this section, δ is a fixed nonnegative number. As well as δ-unimodality, we need to define δ-monotonicity. If $\delta = 0$ then

δ-unimodality and δ-monotonicity reduce to unimodality (Definition 3.1) and monotonicity.

DEFINITION 3.2

Let I be an interval and f a real-valued function on I. We say that f is *strictly δ-monotonic increasing* on I if, for all $x_1, x_2 \in I$,

$$x_1 + \delta < x_2 \supset f(x_1) < f(x_2). \tag{3.11}$$

As an abbreviation, we shall write simply "f is δ-↑ on I". Strictly δ-monotonic decreasing functions (abbreviated δ-↓) are defined in the obvious way.

DEFINITION 3.3

Let I be an interval and f a real-valued function on I. We say that f is *δ-unimodal* on I if, for all $x_0, x_1, x_2 \in I$,

$$x_0 + \delta < x_1 \wedge x_1 + \delta < x_2 \supset (f(x_0) \leq f(x_1) \supset f(x_1) < f(x_2))$$
$$\wedge \, (f(x_1) \geq f(x_2) \supset f(x_0) > f(x_1)). \tag{3.12}$$

The following theorem gives a characterization of δ-unimodal functions. It reduces to Theorem 3.1 if $\delta = 0$.

THEOREM 3.2

f is δ-unimodal on $[a, b]$ iff there exists $\mu \in [a, b]$ such that either f is δ-↓ on $[a, \mu)$ and δ-↑ on $[\mu, b]$, or f is δ-↓ on $[a, \mu]$ and δ-↑ on $(\mu, b]$. Furthermore, if f is δ-unimodal on $[a, b]$, then there is a unique interval $[\mu_1, \mu_2]$ $\subseteq [a, b]$ such that the points μ with the above properties are precisely the elements of $[\mu_1, \mu_2]$, and $\mu_2 \leq \mu_1 + \delta$.

Proof

Suppose μ exists so that f is δ-↓ on $[a, \mu)$ and δ-↑ on $[\mu, b]$. Take any x_0, x_1, x_2 in $[a, b]$ with $x_0 + \delta < x_1$ and $x_1 + \delta < x_2$. If $f(x_0) \leq f(x_1)$ then, since f is δ-↓ on $[a, \mu)$, $\mu \leq x_1$. As f is δ-↑ on $[\mu, b)$, it follows that $f(x_1) < f(x_2)$. The other cases are similar, so f is δ-unimodal.

Conversely, suppose that f is δ-unimodal on $[a, b]$. Let

$$\mu_1 = \inf\{x \in [a, b] \,|\, f \text{ is } \delta\text{-↑ on } [x, b]\}, \tag{3.13}$$

(so $\mu_1 \leq \max(a, b - \delta)$), and

$$\mu_2 = \sup\{x \in [a, b] \,|\, f \text{ is } \delta\text{-↓ on } [a, x]\}, \tag{3.14}$$

(so $\mu_2 \geq \min(a + \delta, b)$).

It is immediate from the definitions (3.13) and (3.14) that f is δ-↑ on $(\mu_1, b]$ and f is δ-↓ on $[a, \mu_2)$. We shall show that

$$\mu_1 \leq \mu_2. \tag{3.15}$$

Suppose, by way of contradiction, that

$$\mu_1 > \mu_2. \tag{3.16}$$

This implies that $\mu_1 > a$ and $\mu_2 < b$. From the definitions of μ_1 and μ_2, there are points x' and x'', with

$$\mu_2 \leq x'' < \left(\frac{\mu_1 + \mu_2}{2}\right) < x' \leq \mu_1, \tag{3.17}$$

such that f is not δ-↑ on $[x', b]$ and f is not δ-↓ on $[a, x'']$. Thus, there are points y', y'', z', z'' in $[a, b]$ such that

$$z'' + \delta < y'' \leq x'' < x' \leq y' < z' - \delta, \tag{3.18}$$

$$f(z'') \leq f(y''), \tag{3.19}$$

and

$$f(y') \geq f(z'). \tag{3.20}$$

Let $x_0 = z''$, $x_2 = z'$, and

$$x_1 = \begin{cases} y' & \text{if } f(y') \geq f(y''), \\ y'' & \text{otherwise.} \end{cases} \tag{3.21}$$

From relations (3.18) to (3.21), the points x_0, x_1, and x_2 contradict δ-unimodality (equation (3.12)). Thus (3.16) is impossible, (3.15) must hold, and $[\mu_1, \mu_2]$ is nonempty.

Choose any μ in $[\mu_1, \mu_2]$. From the definitions of μ_1 and μ_2, f is δ-↓ on $[a, \mu)$ and δ-↑ on $(\mu, b]$. Suppose, if it is possible, that f is neither δ-↓ on $[a, \mu]$ nor δ-↑ on $[\mu, b]$. Then there are points y_1 and y_2, in $[a, b]$, such that

$$y_2 + \delta < \mu < y_1 - \delta, \tag{3.22}$$

$$f(y_1) \leq f(\mu), \tag{3.23}$$

and

$$f(y_2) \leq f(\mu). \tag{3.24}$$

Thus, the points y_2, μ, and y_1 contradict the δ-unimodality of f, so f is either δ-↓ on $[a, \mu]$ or δ-↑ on $[\mu, b]$. This completes the proof of the first part of the theorem.

Finally, by the definitions (3.13) and (3.14), the set of points μ satisfying the conditions of the theorem is precisely $[\mu_1, \mu_2]$. Since f is both δ-↑ and δ-↓ on (μ_1, μ_2), we have $\mu_2 \leq \mu_1 + \delta$, and the proof is complete.

Remarks

The interval $[\mu_1, \mu_2]$ depends on δ. Suppose that f attains its minimum in $[a, b]$ at $\bar{\mu}$. By Theorem 3.2, f is δ-↑ on $(\mu_1, b]$ and δ-↓ on $[a, \mu_2)$, so $\bar{\mu} \in [\mu_2 - \delta, \mu_1 + \delta]$, an interval of length at most 2δ.

As an example, consider

$$f(x) = x^2 + g(x) \tag{3.25}$$

on $[-1, 1]$, where g is any function (not necessarily continuous) with $|g(x)| \leq \epsilon$, and $\epsilon \geq 0$. Since $f(x)$ is bounded above and below by the unimodal functions $x^2 + \epsilon$ and $x^2 - \epsilon$, we see that f is δ-unimodal for any $\delta \geq \sqrt{2\epsilon}$. In a practical case ϵ might be a small multiple of the relative machine precision, and the fact that the least δ for which f is ϵ-unimodal is of order $\epsilon^{1/2}$, rather than ϵ, is to be expected from the discussion in Section 2.

The following theorem is a generalization of Corollary 3.4 (which is just the special case $\delta = 0$), and shows why methods like Fibonacci search and golden section search work on δ-unimodal functions while the distance between points at which f is evaluated is greater than δ.

THEOREM 3.3

Suppose that f is δ-unimodal on $[a, b]$, μ_1 and μ_2 are the points given by Theorem 3.2, x_1 and x_2 are in $[a, b]$, and $x_1 + \delta < x_2$. If $f(x_1) \leq f(x_2)$ then $\mu_2 \leq x_2$, and if $f(x_1) \geq f(x_2)$ then $\mu_1 \geq x_1$.

Proof

If $x_2 < \mu_2$ then $f(x_1) > f(x_2)$ for, by Theorem 3.2 with $\mu = \mu_2$, f is δ-↓ on $[a, \mu_2)$. Hence, if $f(x_1) \leq f(x_2)$ then $\mu_2 \leq x_2$. The second half is similar.

Remarks

Theorems 3.2 and 3.3 show that, provided δ is known, methods like Fibonacci search and golden section search can locate the interval $[\mu_1, \mu_2]$ in an interval of length as close to δ as desired. Since the minimum $\bar{\mu} \in [\mu_2 - \delta, \mu_1 + \delta]$ (see the remarks above), this means that $\bar{\mu}$ can be located in an interval of length as close to 3δ as desired.

In practice f may be δ-unimodal for all $\delta \geq \delta_0$, but a sharp upper bound for δ_0 may be difficult to obtain. If the usual golden section search method is used, giving a nested sequence of intervals I_j with limit $\hat{\mu}$, then Theorem 3.3 shows that $[\mu_1, \mu_2] \subseteq I_j$ as long as the two function evaluations giving I_j were at points separated by more than δ_0. The smallest such interval I_j has length no greater than $(2 + \sqrt{5})\delta_0$, so

$$|\hat{\mu} - \bar{\mu}| \leq (3 + \sqrt{5})\delta_0 \simeq 5.236\delta_0. \tag{3.26}$$

Thus, golden section search gives an approximation $\hat{\mu}$ which is nearly as good as could be expected if we knew δ_0. This may be regarded as a justification for using golden section or Fibonacci search to approximate minima of functions which, because of rounding errors, are only "approximately" unimodal.

Section 4
AN ALGORITHM ANALOGOUS TO DEKKER'S ALGORITHM

For finding a zero of a function f, the bisection process has the advantage that linear convergence is guaranteed, because the interval known to contain a zero is halved at each evaluation of f after the first. However, if f is sufficiently smooth and we have a good initial approximation to a simple zero, then a process with superlinear convergence will be much faster than bisection. This is the motivation for the algorithm, described in Chapter 4, which combines bisection and successive linear interpolation in a way which retains the advantages of both.

There is a clear analogy between methods for finding a minimum and for finding a zero. The Fibonacci and golden section search methods have guaranteed linear convergence, and correspond to bisection. Processes like successive parabolic interpolation, which do not always converge, but under certain conditions converge superlinearly, correspond to successive linear interpolation. In this section we describe an algorithm which combines golden section search and successive parabolic interpolation. The analogy with the algorithm of Chapter 4 is illustrated below.

	Zeros		Extrema
Linear convergence	Bisection	⟷	Golden section search
	↕		↕
Superlinear convergence	Successive linear interpolation	⟷	Successive parabolic interpolation

Many more or less *ad hoc* algorithms have been proposed for one-dimensional minimization, particularly as components of n-dimensional minimization algorithms. See Box, Davies, and Swann (1969); Flanagan, Vitale, and Mendelsohn (1969); Fletcher and Reeves (1964); Jacoby, Kowalik, and Pizzo (1971); Kowalik and Osborne (1968); Pierre (1969); Powell (1964); etc. The algorithm presented here might be regarded as an unwarranted addition to this list, but it seems to be more natural than these algorithms, which involve arbitrary prescriptions like "if . . . fails then halve the step-size and try again". Of course, our algorithm is not quite free of arbitrary prescriptions either; a more objective criticism of the *ad hoc* algorithms is that for many of them convergence to a local minimum in a reasonable number of function evaluations cannot be guaranteed, and, for the exceptions, the asymptotic rate of convergence (when f is sufficiently smooth) is less than for our algorithm (Section 5). Note that we do not claim that our algorithm is suitable for use in an n-dimensional minimization procedure: an *ad hoc* algorithm may be more efficient (see Sections 7.6 and 7.7).

A description of the algorithm

Here we give an outline which should make the main ideas of the algorithm clear. For questions of detail the reader should refer to Section 8, where the algorithm is described formally by the ALGOL 60 procedure *localmin*.

The algorithm finds an approximation to the minimum of a function f defined on the interval $[a, b]$. Unless a is very close to b, f is never evaluated at the endpoints a and b, so f need only be defined on (a, b), and if the minimum is actually at a or b then an interior point distant no more than $2tol$ from a or b will be returned, where tol is a tolerance (see equation (4.2) below). The minimum found may be local, but non-global, unless f is δ-unimodal for some $\delta < tol$.

At a typical step there are six significant points a, b, u, v, w, and x, not all distinct. The positions of these points change during the algorithm, but there should be no confusion if we omit subscripts. Initially (a, b) is the interval on which f is defined, and

$$v = w = x = a + \left(\frac{3 - \sqrt{5}}{2}\right)(b - a). \tag{4.1}$$

The magic number $(3 - \sqrt{5})/2 = 0.381966\ldots$ is rather arbitrarily chosen so that the first step is the same as for a golden section search.

At the start of a cycle (label "loop" of procedure *localmin*) the points a, b, u, v, w, and x always serve as follows: a local minimum lies in $[a, b]$; of all the points at which f has been evaluated, x is the one with the least value of f, or the point of the most recent evaluation if there is a tie; w is the point with the next lowest value of f; v is the previous value of w; and u is the last point at which f has been evaluated (undefined the first time). One possible configuration is shown in Diagram 4.1.

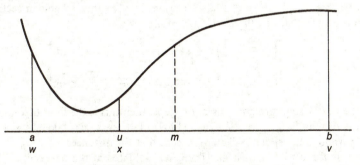

DIAGRAM 4.1 A possible configuration

As in procedure *zero* (Chapter 4), the tolerance is a combination of a relative and an absolute tolerance. If

$$tol = eps\,|x| + t, \tag{4.2}$$

then the point x returned approximates a minimum to an accuracy of $2tol + \delta < 3tol$, provided f is δ-unimodal near x and $\delta < tol$. The user must provide the positive parameters *eps* and *t*. In view of the discussion in Section 2, it is generally unreasonable to take *eps* much less than $\epsilon^{1/2}$, where ϵ is the machine-precision (see Section 4.2). The parameter *t* should be positive in case the minimum is at 0. It is possible that the error may exceed $2tol + \delta$ because of the effect of rounding errors in determining if the stopping criterion is satisfied, but the additional error is negligible if *eps* is of order $\epsilon^{1/2}$ or greater.

Let $m = \frac{1}{2}(a + b)$ be the midpoint of the interval known to contain the minimum. If $|x - m| \leq 2tol - \frac{1}{2}(b - a)$, i.e., if $\max(x - a, b - x) \leq 2tol$, then the procedure terminates with x as the approximate position of the minimum. Otherwise, numbers p and $q(q \geq 0)$ are computed so that $x + p/q$ is the turning point of the parabola passing through $(v, f(v))$, $(w, f(w))$, and $(x, f(x))$. If two or more of these points coincide, or if the parabola degenerates to a straight line, then $q = 0$.

p and q are given by

$$p = \pm[(x - v)^2(f(x) - f(w)) - (x - w)^2(f(x) - f(v))] \qquad (4.3)$$

$$= \pm(x - v)(x - w)(w - v)\{(x - w)f[v, w, x] + f[w, x]\}, \qquad (4.4)$$

and

$$q = \mp 2[(x - v)(f(x) - f(w)) - (x - w)(f(x) - f(v))] \qquad (4.5)$$

$$= \mp 2(x - v)(x - w)(w - v)f[v, w, x]. \qquad (4.6)$$

From (4.4) and (4.6), the correction p/q should be small if x is close to a minimum where the second derivative is positive, so the effect of rounding errors in computing p and q is minimized. (Golub and Smith (1967) compute a correction to $\frac{1}{2}(v + w)$ for the same reason.)

As in procedure *zero*, let e be the value of p/q at the second-last cycle. If $|e| \leq tol$, $q = 0$, $x + p/q \notin (a, b)$, or $|p/q| \geq \frac{1}{2}|e|$, then a "golden section" step is performed, i.e., the next value of u is

$$u = \begin{cases} \left(\frac{\sqrt{5} - 1}{2}\right)x + \left(\frac{3 - \sqrt{5}}{2}\right)a & \text{if } x \geq m, \\ \left(\frac{\sqrt{5} - 1}{2}\right)x + \left(\frac{3 - \sqrt{5}}{2}\right)b & \text{if } x < m. \end{cases} \qquad (4.7)$$

(If the next k steps are golden section steps, then this is the limit of the optimal choice as $k \to \infty$: see Witzgall (1969).) Otherwise u is taken as $x + p/q$ (a "parabolic interpolation" step), except that the distances $|u - x|$, $u - a$, and $b - u$ must be at least *tol*. Then f is evaluated at the new point u, the points a, b, v, w, and x are updated as necessary, and the cycle is repeated (the procedure returns to the label "loop"). We see that f is never evaluated at two points closer together than *tol*, so δ-unimodality for some $\delta < tol$ is enough to ensure that the global minimum is found to an accuracy of $2tol + \delta$ (see Theorem 3.3 and the following remarks).

Typically the algorithm terminates in the following way: $x = b - tol$ (or, symmetrically, $a + tol$) after a parabolic interpolation step has been performed with the condition $|u - x| \geq tol$ enforced. The next parabolic interpolation point lies very close to x and b, so u is forced to be $x - tol$. If $f(u) > f(x)$ then a moves to u, $b - a$ becomes $2tol$, and the termination criterion is satisfied (see Diagram 4.2). Note that two consecutive steps of *tol* are done just before termination. If a golden section search were done whenever the last, rather than second-last, value of $|p/q|$ was *tol* or less, then termination with two consecutive steps of *tol* would be prevented, and unnecessary golden section steps would be performed.

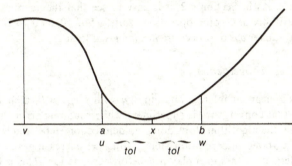

DIAGRAM 4.2   **A typical configuration after termination**

Section 5
CONVERGENCE PROPERTIES

There cannot be more than about $2\log_2 [(b - a)/tol]$ consecutive parabolic interpolation steps (with the current a and b, and the minimum of *tol* over the interval), for while parabolic interpolation steps are being performed $|p/q|$ decreases by a factor of at least two on every second cycle of the algorithm, and when $|e| \leq tol$ a golden section step is performed. (In this section, "about" means we are not distinguishing between a real number and its integer part. Precise results may easily be obtained as in Section 4.3.) A golden section step does not necessarily decrease $b - a$ significantly, e.g., if $x = b - tol$ and $f(u) < f(x)$, then $b - a$ is only decreased by *tol*, but two golden section steps must decrease $b - a$ by a factor of at least $(1 + \sqrt{5})/2$ $= 1.618\ldots$ As in Section 4.3, we see that convergence cannot require more than about

$$2K\left[\log_2\left(\frac{b - a}{tol}\right) \right]^2 \tag{5.1}$$

function evaluations, where

$$K = \frac{1}{\log_2[(1 + \sqrt{5})/2]} = 1.44\ldots. \tag{5.2}$$

By comparison, a golden section or Fibonacci search would require about

$$K \log_2\left(\frac{b-a}{tol}\right) \tag{5.3}$$

function evaluations, and a brute-force search about $(b - a)/(2tol)$.

The analogy with procedure *zero* of Chapter 4 should be clear, and essentially the same remarks apply here as were made in Section 4.3. In practical tests convergence has never been more than 5 percent slower than for a Fibonacci search (see Section 6).

In deriving (5.1) we have ignored the effect of rounding errors inside the procedure. As in Section 4.2, it is easy to see that they cannot prevent convergence if floating-point operations satisfy (4.2.10) and (4.2.11), provided the parameter *eps* of procedure *localmin* is at least 2ϵ.

Superlinear convergence

If f is C^2 near an interior minimum μ with $f''(\mu) > 0$, then Theorem 3.4.1 shows that convergence is superlinear while rounding errors are negligible. Usually the algorithm stops doing golden section steps, and eventually does only parabolic interpolation steps, with $f(x)$ decreasing at each step, until the tolerance comes into play just before termination. This is certainly true if the successive parabolic interpolation process converges with strong order $\beta_2 = 1.3247 \ldots$ (sufficient conditions for this are given in Sections 3.6 and 3.7).

For most of the *ad hoc* methods given in the literature, convergence with a guaranteed error bound of order *tol* in the number of steps given by (5.1) is not certain, and, even if convergence does occur, the order is no greater than for our algorithm. For example, the algorithm of Davies, Swann, and Campey (Box, Davies, and Swann (1969)) evaluates f at two or more points for each parabolic fit, so the order of convergence is at most $\sqrt{\beta_2} = 1.150 \ldots$.

Section 6
PRACTICAL TESTS

The ALGOL procedure *localmin* given in Section 8 has been tested using ALGOL W (Wirth and Hoare (1966); Bauer, Becker, and Graham (1968)) on IBM 360/67 and 360/91 computers with machine precision 16^{-13}. Although it is possible to contrive an example where the bound (5.1) on the number of function evaluations is nearly attained, for our test cases convergence requires, at worst, only 5 percent more function evaluations than are needed to guarantee the same accuracy using Fibonacci search. In most practical cases superlinear convergence sets in after a few golden section steps, and the procedure is much faster than Fibonacci search.

As an example, in Table 6.1 we give the number of function evaluations required to find the minima of the function

$$f(x) = \sum_{i=1}^{20} \left(\frac{2i-5}{x-i^2}\right)^2. \tag{6.1}$$

This function has poles at $x = 1^2, 2^2, \ldots, 20^2$. Restricted to the open interval $(i^2, (i+1)^2)$ for $i = 1, 2, \ldots, 19$, it is unimodal (ignoring rounding errors) with an interior minimum. The fourth column of Table 6.1 gives the number n_L of function evaluations required to find this minimum μ_i, using procedure *localmin* with $eps = 16^{-7}$ and $t = 10^{-10}$ (so the error bound is less than $3tol$, where $tol = 16^{-7}|\mu_i| + 10^{-10}$).

The last column of the table gives the number n_Z of function evaluations required to find the zero of

$$f'(x) = -2 \sum_{i=1}^{20} \frac{(2i-5)^2}{(x-i^2)^3} \tag{6.2}$$

in the interval $[i^2 + 10^{-9}, (i+1)^2 - 10^{-9}]$, using procedure *zero* (Section 4.6) with *macheps* $= 16^{-7}$ and $t = 10^{-10}$, so the guaranteed accuracy is nearly the same as for *localmin*. Of course, in practical cases we would seldom be lucky enough to have such a simple analytic expression for f', so procedure

TABLE 6.1   **Comparison of procedures *localmin* and *zero***

i	μ_i	$f(\mu_i)$	n_L	n_Z
1	3.0229153	3.6766990169	12	14
2	6.6837536	1.1118500100	11	8
3	11.2387017	1.2182217637	13	14
4	19.6760001	2.1621103109	10	12
5	29.8282273	3.0322905193	11	12
6	41.9061162	3.7583856477	11	11
7	55.9535958	4.3554103836	10	11
8	71.9856656	4.8482959563	10	11
9	90.0088685	5.2587585400	10	10
10	110.0265327	5.6036524295	10	10
11	132.0405517	5.8956037976	10	10
12	156.0521144	6.1438861542	9	10
13	182.0620604	6.3550764593	9	10
14	210.0711010	6.5333662003	9	10
15	240.0800483	6.6803639849	9	10
16	272.0902669	6.7938538365	9	10
17	306.1051233	6.8634981053	9	10
18	342.1369454	6.8539024631	9	9
19	380.2687097	6.6008470481	9	9

zero could not easily be used to find minima of *f* in this manner. Also, procedure *zero* could find a maximum rather than a minimum.

Table 6.1 shows that the number of function evaluations required by procedure *localmin* compares favorably with the number required by procedure *zero*. Both are much faster than Fibonacci search, which would require 45 function evaluations to find the minimum for $i = 10$ to the same accuracy.

For some numerical results illustrating the superlinear convergence of the successive parabolic interpolation process, see Section 3.9.

Section 7
CONCLUSION

The algorithm given in this chapter has the same advantages as the algorithm described in Chapter 4 for finding zeros: convergence in a reasonable number of steps is guaranteed for any function (see equation (5.1)), and on well-behaved functions convergence is superlinear, with order at least 1.3247 ..., and thus much faster than Fibonacci search. There is no contradiction here: Fibonacci search is the fastest method for the worst possible function, but our algorithm is faster on a large class of functions, including, for example, C^2 functions with positive second derivatives at interior minima.

A similar algorithm using derivatives

We pointed out in Section 4.5 that bisection could be combined with interpolation formulas which use both *f* and *f'*. We could combine golden section search with an interpolation method using both *f* and *f'* in a similar way. Davidon (1959) suggests fitting a cubic polynomial to agree with *f* and *f'* at two points, and taking a turning point of the cubic as the next approximation. (See also Johnson and Myers (1967).) This method, which gives the possibility of superlinear convergence, could well replace successive parabolic interpolation (using *f* at three points) in our algorithm if *f'* is easy to compute. If the cubic has no real turning point, or if the turning point which is a local minimum lies outside the interval known to contain a minimum of *f*, then we can resort to golden section search.

Parallel algorithms

So far we have considered only serial (i.e., sequential) algorithms for finding minima. If a parallel computer is available, more efficient algorithms which take advantage of the parallelism are possible, just as in the analogous zero-finding problem (see Section 4.5). Karp and Miranker (1968) give a parallel search method which is a generalization of Fibonacci search, and optimal in the same sense, if a sufficiently parallel processor is available. See also Wilde (1964) and Avriel and Wilde (1966). Miranker (1969) gives parallel methods for approximating the root of a function, and these could be used to find a root of f'. (Parallel methods for finding a root of f', using only evaluations of f, could also be used.) These parallel methods could be combined to give a parallel method with guaranteed convergence, and often superlinear convergence with a higher order than for our serial method.

Section 8
AN ALGOL 60 PROCEDURE

The ALGOL procedure *localmin* for finding a local minimum of a function of one variable is given below. The algorithm and some numerical results are described in Sections 4 to 6. A FORTRAN translation of procedure *localmin* is given in the Appendix.

real procedure *localmin* (a, b, eps, t, f, x);
value a, b, eps, t; **real** a, b, eps, t, x; **real procedure** f;
   begin comment:
      If the function f is defined on the interval (a, b), then *localmin* finds an approximation x to the point at which f attains its minimum (or the appropriate limit point), and returns the value of f at x. t and *eps* define a tolerance $tol = eps\,|x| + t$, and f is never evaluated at two points closer together than *tol*. If f is δ-unimodal (Definition 3.3) for some $\delta < tol$, then x approximates the global minimum of f with an error less than $3tol$ (see Section 4). If f is not δ-unimodal on (a, b), then x may approximate a local, but non-global, minimum. *eps* should be no smaller than $2macheps$, and preferably not much less than sqrt (*macheps*), where *macheps* is the relative machine precision (Section 4.2). t should be positive. For further details, see Section 2.
      The method used is a combination of golden section search and successive parabolic interpolation. Convergence is never much slower than for a Fibonacci search (see Sections 5 and 6). If f has a continuous second derivative which is positive at the minimum (not at a or b) then, ignoring rounding errors, convergence is superlinear, and usually the order is at least $1.3247\ldots$;
   real $c, d, e, m, p, q, r, tol, t2, u, v, w, fu, fv, fw, fx$;
   $c := 0.381966$; **comment**: $c = (3 - \text{sqrt}(5))/2$;
   $v := w := x := a + c \times (b - a); e := 0$;
   $fv := fw := fx := f(x)$;
   comment: Main loop;
   loop: $m := 0.5 \times (a + b)$;
   $tol := eps \times \text{abs}(x) + t; t2 := 2 \times tol$;

comment: Check stopping criterion;
if abs$(x - m) > t2 - 0.5 \times (b - a)$ **then**

    begin $p: = q: = r: = 0$;

    if abs$(e) > tol$ **then**

        begin comment: Fit parabola;

        $r: = (x - w) \times (fx - fv); q: = (x - v) \times (fx - fw);$

        $p: = (x - v) \times q - (x - w) \times r; q: = 2 \times (q - r);$

        if $q > 0$ **then** $p: = -p$ **else** $q: = -q$;

        $r: = e; e: = d$

        end;

    if abs$(p) < $ abs$(0.5 \times q \times r) \land p < q \times (a - x) \land$

        $p < q \times (b - x)$ **then**

    begin comment: A "parabolic interpolation" step;

    $d: = p/q; u: = x + d;$

    comment: f must not be evaluated too close to a or b;

    if $u - a < t2 \lor b - u < t2$ **then** $d: = $ **if** $x < m$ **then** tol

      else $-tol$

    end

    else

        begin comment: A "golden section" step;

        $e: = ($ **if** $x < m$ **then** b **else** $a) - x; d: = c \times e$

        end;

    comment: f must not be evaluated too close to x;

    $u: = x + ($ **if** abs$(d) \geq tol$ **then** d **else if** $d > 0$ **then** tol **else** $-tol)$;

    $fu: = f(u)$;

    comment: Update a, b, v, w, and x;

    if $fu \leq fx$ **then**

        begin if $u < x$ **then** $b: = x$ **else** $a: = x$;

        $v: = w; fv: = fw; w: = x; fw: = fx; x: = u; fx: = fu$

        end

    else

        begin if $u < x$ **then** $a: = u$ **else** $b: = u$;

        if $fu \leq fw \lor w = x$ **then**

            begin $v: = w; fv: = fw; w: = u; fw: = fu$ **end**

        else if $fu \leq fv \lor v = x \lor v = w$ **then**

            begin $v: = u; fv: = fu$

            end

        end;

    go to loop

    end;

$localmin: = fx$

end $localmin$;

6

GLOBAL MINIMIZATION
GIVEN AN UPPER BOUND
ON THE SECOND DERIVATIVE

Section 1
INTRODUCTION

Minimization procedures like the one described in Chapter 5 can only guarantee to find a local, not necessarily global, minimum of a function $f \in C[a, b]$. If f happens to be unimodal then a local minimum must be the global minimum, but in practical problems it often happens that f is not unimodal, or that unimodality is difficult to prove. In this chapter we investigate the problem of finding a good approximation to the global minimum, given weaker conditions on f than unimodality. As usual, we consider methods which depend on the sequential evaluation of f at a finite number of points, and our aim is to reduce, as far as possible, the number of function evaluations required to give an answer which is guaranteed to be accurate to within some prescribed tolerance.

In Sections 2 to 6 we describe an efficient algorithm for approximating the global minimum of a function of one variable, given an upper bound on the second derivative. There are many obvious applications for this algorithm. For example, when finding *a posteriori* error bounds for the approximate solution of elliptic partial differential equations, we may need to find the maximum of $|f(x)|$ (Fox, Henrici, and Moler (1967)). Instead of working with $|f(x)|$, which may have discontinuous derivatives, it is probably better to use the relation

$$\max_x |f(x)| = -\min[\min_x f(x), \min_x(-f(x))]. \tag{1.1}$$

In Sections 7 and 8 we show how to extend the method to functions of several variables, and ALGOL 60 procedures are given in Section 10.

Some fundamental limitations

If $f \in C[a, b]$, let

$$\varphi_f = \inf \{f(x) \,|\, x \in [a, b]\} \tag{1.2}$$

and

$$\mu_f = \inf \{x \in [a, b] \,|\, f(x) = \varphi_f\}. \tag{1.3}$$

Even if f satisfies very stringent smoothness conditions, the problem of finding μ_f is improperly posed, in the sense that μ_f is not a continuous function of f (with the uniform topology on $C[a, b]$). For example, consider

$$f_\delta(x) = \cos(\pi x) - \delta x \tag{1.4}$$

on $[-2, 2]$. If $\delta > 0$ then $\mu_f \simeq 1$, but if $\delta \leq 0$ then $\mu_f \simeq -1$, so a very small change in f can cause a large change in μ_f.

Instead of trying to approximate μ_f, we should seek to approximate $\varphi_f = f(\mu_f)$. Since

$$|\varphi_f - \varphi_g| \leq \|f - g\|_\infty \tag{1.5}$$

for all f and g in $C[a, b]$, φ is a continuous function on $C[a, b]$, so the problem of finding φ_f is properly posed. However, given $t > 0$, it is still impossible to find $\hat{\varphi}$ such that

$$|\hat{\varphi} - \varphi_f| \leq t \tag{1.6}$$

with a finite number N_t of function evaluations, unless we have some *a priori* information about f.

A priori *conditions on f*

If $f \in C[a, b]$, the modulus of continuity $w(f; \delta)$ is defined (as in Section 2.2) by

$$w(f; \delta) = \sup_{\substack{|x-y| \leq \delta \\ x,y \in [a,b]}} |f(x) - f(y)| \tag{1.7}$$

for $\delta \geq 0$. Suppose that a function $W(\delta)$ is given such that

$$\lim_{\delta \to 0+} W(\delta) = 0, \tag{1.8}$$

and

$$w(f; \delta) \leq W(\delta) \tag{1.9}$$

for all $\delta > 0$. Given $t > 0$, choose $\delta > 0$ such that

$$W(\delta) \leq t \tag{1.10}$$

(always possible by (1.8)), and evaluate f at points $x_0, \ldots, x_n$ in $[a, b]$ such that

$$\max_{x \in [a,b]} \min_{0 \le i \le n} |x - x_i| \le \delta. \tag{1.11}$$

(For example, we might choose $x_0 = a + \delta$, $x_1 = a + 3\delta$, $x_2 = a + 5\delta$, etc.) If

$$\hat{\varphi} = \min_{0 \le i \le n} f(x_i) \tag{1.12}$$

then, from (1.7), (1.9), (1.10), and (1.11),

$$0 \le \hat{\varphi} - \varphi_f \le t. \tag{1.13}$$

Thus, a quite weak condition enabling us to approximate φ_f with a finite number of function evaluations is that we have a bound $W(\delta)$, satisfying (1.8), on the modulus of continuity $w(f; \delta)$ of f.

For example, if $f \in C^1[a, b]$ and

$$\|f'\|_\infty \le M, \tag{1.14}$$

then we can take

$$W(\delta) = M\delta. \tag{1.15}$$

The procedure suggested above will be very slow if t is small: in fact, about $(b - a)M/(2t)$ function evaluations will be required. However, it may be impossible to do much better than this without knowing more about f. Consider minimizing a function which is known to be in the class

$$\{f_c(x) = \min (1.01t, M|x - c|) \mid c \in [a, b]\}. \tag{1.16}$$

If

$$\delta = \frac{1.01t}{M}, \tag{1.17}$$

and $\hat{\varphi}$ is computed from (1.12) for some set of points $x_0, \ldots, x_n$, then there is a choice of $c \in [a, b]$ for which $\hat{\varphi}$ fails to satisfy (1.13) unless (1.11) holds, so at least $\lceil (b - a)M/(2.02t) \rceil$ function evaluations are required. Sometimes fewer function evaluations are necessary: for example, if

$$f(x) = Mx, \tag{1.18}$$

then it is enough to evaluate f at a and b. (See also Section 5.)

Instead of having an *a priori* bound on $\|f'\|_\infty$, we could have a bound

$$\|f^{(r)}\|_\infty \le M \tag{1.19}$$

on $\|f^{(r)}\|_\infty$, for some $r \ge 1$. We show below that, with such a bound, the maximum number of function evaluations required to find $\hat{\varphi}$ satisfying (1.13) is of order $(M/t)^{1/r}$.

The case $r = 1$ is discussed above, so suppose $r \ge 2$, and let

$$n = \left\lceil \frac{(b - a)}{4 \cos \left(\dfrac{\pi}{2r}\right)} \left(\frac{4M}{r!\, t}\right)^{1/r} \right\rceil. \tag{1.20}$$

Define $\delta = (b - a)/n$, $a_i = a + i\delta$ for $i = 0, \ldots, n$ (so $a_n = b$), and

$$a_{i,j} = a_i + \frac{\delta}{2}\left\{1 - \frac{\cos\left(\dfrac{(2j-1)\pi}{2r}\right)}{\cos\left(\dfrac{\pi}{2r}\right)}\right\} \tag{1.21}$$

for $i = 0, \ldots, n - 1$ and $j = 1, \ldots, r$ (so $a_{i,1} = a_i$, $a_{i,r} = a_{i+1}$). Let $P_i = IP(f; a_{i,1}, \ldots, a_{i,r})$ be the polynomial of degree $r - 1$ which coincides with f at $a_{i,1}, \ldots, a_{i,r}$. Lemma 2.4.1 and the bound (1.19) show that, for all $x \in [a_i, a_{i+1}]$,

$$|f(x) - P_i(x)| \leq |(x - a_{i,1}) \ldots (x - a_{i,r})| \frac{M}{r!}. \tag{1.22}$$

The right side of (1.22) is no greater than $\{\delta/[2\cos(\pi/2r)]\}^r M/(r!\, 2^{r-1})$ and, by (1.20) and the choice of δ, this is no greater than $t/2$. Thus, we need only find the minimum of each polynomial $P_i(x)$ in $[a_i, a_{i+1}]$ to within a tolerance $t/2$. This is easy if $r = 2$, for then each polynomial $P_i(x)$ is linear. If $r > 2$, then we can bound $|P_i'(x)|$ in $[a_i, a_{i+1}]$, and apply the procedure for $r = 2$ to minimize $P_i(x)$. (This idea for finding bounds on polynomials in an interval was suggested by Rivlin (1970). Another possibility is to minimize $P_i(x)$ by the method of Goldstein and Price (1971).) Because successive intervals $[a_i, a_{i+1}]$ are adjacent, the number of function evaluations required to find $\hat{\phi}$ satisfying (1.13) does not exceed

$$N = (r - 1)n + 2, \tag{1.23}$$

where n is given by (1.20).

Since N is of order $(M/t)^{1/r}$, the method described above is not likely to be practical for small t unless $r \geq 2$. On the other hand, in practical problems it is usually difficult to obtain good bounds on the third or higher derivatives of f (if they exist). Thus, in the rest of this chapter we suppose that $r = 2$. It turns out that a one-sided bound

$$f''(x) \leq M \tag{1.24}$$

is sufficient, instead of the two-sided bound (1.19). If $f''(x)$ has a physical interpretation (e.g., as an acceleration), then a bound of the form (1.24) can sometimes be obtained from physical considerations.

Section 2
THE BASIC THEOREMS

The global minimization algorithm which is described in the next section depends on the simple Theorems 2.1, 2.2, and 2.3. Theorem 2.1 is related to the maximum principle for elliptic difference operators, and also to some results in Davis (1965). We assume that $f \in C^1[a, b]$, and that

$$f'(x) - f'(y) \leq M(x - y) \tag{2.1}$$

for all x, y in $[a, b]$ with $x > y$. (Weaker conditions suffice: see Section 7.)
If $f \in C^2[a, b]$, then the one-sided Lipschitz condition (2.1) is equivalent to

$$f''(x) \leq M \tag{2.2}$$

for all $x \in [a, b]$.

THEOREM 2.1

Suppose (2.1) holds. Then, for all $x \in [a, b]$,

$$f(x) \geq \frac{(b - x)f(a) + (x - a)f(b)}{b - a} - \frac{1}{2}M(x - a)(b - x). \tag{2.3}$$

The proof is immediate from Lemma 2.4.1.

LEMMA 2.1

Suppose (2.1) holds and $a < 0 \leq b$. Then

$$f'(0) \leq \frac{f(a) - f(0)}{a} - \frac{1}{2}Ma. \tag{2.4}$$

Proof

Applying Lemma 2.3.1 to $f(-x)$, we have

$$f(a) \leq f(0) + af'(0) + \tfrac{1}{2}Ma^2, \tag{2.5}$$

so the result follows.

THEOREM 2.2

Suppose (2.1) holds, $M > 0$, $a < c \leq b$, $f(a) \geq f(c)$, and $f'(c) = 0$.
Then

$$c - a \geq \sqrt{\frac{f(a) - f(c)}{\tfrac{1}{2}M}}. \tag{2.6}$$

Proof

Applying Lemma 2.1 with a suitable translation of the origin gives

$$0 = f'(c) \leq \frac{f(a) - f(c)}{a - c} - \frac{1}{2}M(a - c), \tag{2.7}$$

so

$$f(a) - f(c) \leq \tfrac{1}{2}M(c - a)^2, \tag{2.8}$$

and the result follows.

LEMMA 2.2

Suppose (2.1) holds, $M > 0$, and $a < 0 \leq b \leq -f'(0)/M$. Then $f'(b) \leq 0$.

Proof

By condition (2.1),

$$f'(b) \leq f'(0) + Mb, \tag{2.9}$$

but

$$b \leq \frac{-f'(0)}{M},$$ (2.10)

so the result follows.

THEOREM 2.3

Suppose (2.1) holds, $M > 0$, $a < c \leq b$, and

$$c \leq x \leq \min\left(b, \frac{a+c}{2} - \frac{f(a)-f(c)}{M(a-c)}\right).$$ (2.11)

Then

$$f'(x) \leq 0.$$ (2.12)

Proof

There is no loss of generality in assuming that $c = 0$ and $b = x$. By condition (2.11),

$$b = x \leq \frac{1}{2}a - \frac{f(a)-f(0)}{Ma} = -\frac{1}{M}\left(\frac{f(a)-f(0)}{a} - \frac{1}{2}Ma\right),$$ (2.13)

so, by Lemma 2.1, we have

$$b \leq \frac{-f'(0)}{M}.$$ (2.14)

Now the result follows from Lemma 2.2.

Remarks

Theorems 2.1, 2.2, and 2.3 are sharp, as can easily be seen by taking $f(x)$ as a suitable parabola with leading term $\frac{1}{2}Mx^2$. The theorems are generalized in Section 7, and the proofs given there show that everything needed to justify our minimization algorithm follows from the fundamental inequality (2.3). The proofs given in this section are, however, simpler and more intuitive than those in Section 7.

Section 3
AN ALGORITHM FOR GLOBAL MINIMIZATION

Suppose that $f \in C^2[a, b]$ and, for all $x \in [a, b]$,

$$f''(x) \leq M.$$ (3.1)

We want to find $\hat{\mu} \in [a, b]$ and $\hat{\varphi} = f(\hat{\mu})$ satisfying

$$|\hat{\varphi} - \varphi_f| \leq t,$$ (3.2)

where t is a given positive tolerance and

$$\varphi_f = \min_{x \in [a,b]} f(x).$$ (3.3)

If $M \leq 0$ the problem is quite trivial, for Theorem 2.1 says that $f(x)$ cannot lie below the straight line interpolating f at a and b, so

$$\varphi_f = \min\left(f(a), f(b)\right). \tag{3.4}$$

If $M > 0$ the problem is not trivial, although we saw in Section 1 that there does exist an algorithm to solve it.

The basic algorithm

The algorithm described in this section is an elaboration and refinement of the following basic algorithm. (The notation is consistent with that of the ALGOL procedure *glomin* (Section 10), except that we write M for m, $\hat{\mu}$ for x, $\hat{\varphi}$ for y ($= glomin$), and ϵ for *macheps*.)

1. Set $\hat{\varphi} \leftarrow \min\left(f(a), f(b)\right)$, $\hat{\mu} \leftarrow$ if $\hat{\varphi} = f(a)$ then a else b, and $a_2 \leftarrow a$.
2. If $M \leq 0$ or $a_2 \geq b$ then halt. Otherwise set $a_3 \leftarrow$ some point in $(a_2, b]$ (e.g., b: see below for a better choice).
3. If $f(a_3) < \hat{\varphi}$ then set $\hat{\mu} \leftarrow a_3$ and $\hat{\varphi} \leftarrow f(a_3)$.
4. If the parabola $y = P(x)$, with $P''(x) = M$, $P(a_2) = f(a_2)$, and $P(a_3) = f(a_3)$, satisfies $P(x) \geq \hat{\varphi} - t$ for all x in $[a_2, a_3]$, then go to 5. Otherwise set $a_3 \leftarrow \frac{1}{2}(a_2 + a_3)$ and go back to 3.
5. Set $a_2 \leftarrow a_3$ and go back to 2.

We shall see shortly that (with a sensible choice of a_3 at step 2) the basic algorithm must terminate in a finite number of steps. In view of Theorem 2.1 and step 4, it is clear that the algorithm terminates with $\hat{\varphi}$ satisfying (3.2).

Refinements of the basic algorithm

The crux of the problem is how to make a good choice of a_3 at step 2 of the basic algorithm. We want to choose a_3 as large as possible, but not so large that it has to be reduced at step 4. Theorems 2.2 and 2.3 provide useful lower bounds. If the global minimum μ_f lies outside (a_2, b), or if $\varphi_f \geq \hat{\varphi} - t$, then we may halt, for $\hat{\varphi}$ already satisfies (3.2). Otherwise

$$f'(\mu_f) = 0 \tag{3.5}$$

and

$$f(\mu_f) < \hat{\varphi} - t, \tag{3.6}$$

so, from Theorem 2.2 with a replaced by a_2 and c by μ_f,

$$\mu_f - a_2 > \sqrt{\frac{f(a_2) - \hat{\varphi} + t}{\frac{1}{2}M}}. \tag{3.7}$$

Thus, at step 2 it is safe to take $a_3 = a_3'$, where

$$a_3' = \min\left\{b, a_2 + \sqrt{\frac{f(a_2) - \hat{\varphi} + t}{\frac{1}{2}M}}\right\}, \tag{3.8}$$

and with this choice there is no risk that a_3 will have to be reduced at step 4. Since the right side of (3.7) is at least $(2t/M)^{1/2}$, the basic algorithm must converge in a finite number of steps if, at step 2, we choose any a_3 in the range $[a_3', b]$.

If f is decreasing rapidly at a_2, then Theorem 2.3 may give a better bound than (3.7). Apply Theorem 2.3 with c replaced by a_2 and a replaced by a point $a_2 - d_0$ (with $d_0 > 0$) where f has already been evaluated. (This is not possible if $a_2 = a$.) Combining the result with (3.8), we see that it is safe to choose $a_3 = a_3''$ at step 2, where

$$a_3'' = \min\left(b, \max\left\{a_2 + \sqrt{\frac{f(a_2) - f(\hat{\mu}) + t}{\tfrac{1}{2}M}}\right.\right.,$$
$$\left.\left. a_2 - \frac{1}{2}\left(d_0 + \frac{f(a_2) - f(a_2 - d_0) + 2.01e}{\tfrac{1}{2}Md_0}\right)\right\}\right). \tag{3.9}$$

Here e is a positive tolerance, and the term $2.01e$ is introduced to combat the effect of rounding errors (see equations (3.41) and (3.52)).

The choice $a_3 = a_3''$ is safe, but it is possible to speed up the algorithm by sometimes choosing $a_3 > a_3''$. Because we want to avoid having to decrease a_3 at step 4, the best choice would be to take $a_3 = \min(b, a_3^*)$ where a_3^* is the abscissa of the point to the right of a_2 where the curve $y = f(x)$ intersects the parabola P, with second derivative M, which passes through $(a_2, f(a_2))$ and attains its minimum value $\varphi' - t$ to the right of a_2. Here

$$\varphi' = \min(\hat{\varphi}, f(a_3)) \tag{3.10}$$

is the value of $\hat{\varphi}$ after step 3 has been executed, and we can extend the domain of f by defining $f(x) = f(b)$ for $x > b$ if this is necessary. A typical situation is illustrated in Diagram 3.1.

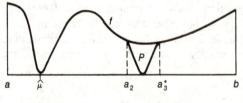

DIAGRAM 3.1   A typical situation

It is not practical to choose $a_3 = a_3^*$, for, although a_3^* exists, several function evaluations are needed to approximate it accurately. Procedure *glomin* (Section 10) finds a rough approximation a_3^{**} to a_3^*, without any extra function evaluations, by assuming that f can be approximated sufficiently well by the parabola which interpolates f at the last three points at which f has been evaluated. To avoid overstepping a_3^* too often because of the inadequacy of the parabolic approximation to f, the procedure uses a heuristic "safety factor" $h \in (0, 1)$. If

$$\hat{a}_3 = \min (b, a_2 + h(a_3^{**} - a_2)), \qquad (3.11)$$

then at step 2 we choose

$$a_3 = \max (a_3'', \hat{a}_3), \qquad (3.12)$$

and if it is neccessary to reduce a_3 at step 4 then we set $a_3 \leftarrow \max (a_3'', \frac{1}{2}(a_2 + a_3))$. Procedure *glomin* also makes a rather primitive attempt to adjust h, the adjustment depending on the outcome of step 4.

Some details of procedure glomin

The ALGOL 60 procedure *glomin* given in Section 10 uses the basic algorithm with the refinements suggested above. From equation (3.8) and the criterion in step 4 of the basic algorithm it is clear that, to speed up convergence, we want to find a rough approximation to the global minimum as soon as possible. In other words, $\hat{\varphi}$ should be nearly at its final value as soon as possible. For this reason, procedure *glomin* incorporates several heuristic strategies which are designed to reduce $\hat{\varphi}$ quickly. We emphasize that the global minimum would be found without using these strategies: the strategies merely reduce the number of function evaluations required (see Sections 5 and 6).

The first strategy for reducing $\hat{\varphi}$ quickly is a pseudo-random search. About ten percent of the function evaluations are used to evaluate f at "random" points uniformly distributed in (a_2, b). (f is not evaluated at the random point a_3 if Theorem 2.1, with a replaced by a_2 and x by a_3, indicates that $f(a_3) \geq \hat{\varphi} - t$, for such an evaluation would be a waste of time.) At worst, this strategy wastes ten percent of the function evaluations, but the saving in function evaluations caused by quickly finding a good value of $\hat{\varphi}$ is usually much more than ten percent. The arbitrary choice of ten percent was made after some numerical experiments.

By comparison with the random search strategy, the second strategy is highly nonrandom. f is evaluated at the minimum of the parabola which interpolates f at the last three points at which f has been evaluated, provided that this minimum a_3 lies in (a_2, b) and Theorem 2.1 does not show that the evaluation is futile for the purpose of reducing $\hat{\varphi}$. The details are similar to those of procedure *localmin* (see Chapter 5). This strategy helps to locate the local minima of f which are in the interior of $[a, b]$, and, unless the global minimum is at a or b, one of these local minima is the global minimum. A bonus is that, if f is sufficiently well-behaved near the global minimum (see Chapter 3 for more precise conditions), then the minimum will be found more accurately than would be expected with the basic algorithm. The numer-

ical examples given in Sections 6 and 8 illustrate this. To avoid wasting function evaluations by repeatedly finding the same local minimum, this strategy
is only used once in about every tenth cycle, although it is always used if
$\hat{\varphi} = f(a_2)$, for then there is a good chance that $f(a_3) < \hat{\varphi}$.

Finally, the user may be able to make a good guess at the global minimum. For example, he may know a local minimum which is likely to be the
global minimum, or he may know the global minimum of a slightly different
function (see the application discussed in Section 8). Thus, procedure *glomin*
has an input parameter c which may be set by the user at the suspected position of the global minimum, and on entry the procedure evaluates f at c in
an attempt to reduce $\hat{\varphi}$. If the user knows nothing about the likely position
of the global minimum, he can set $c = a$ or b.

We can now summarize procedure *glomin*. (For points of detail, see
Section 10.) Step 1 of the basic algorithm is performed, and the algorithm
terminates immediately unless $M > 0$ and $a < b$. Before choosing
$a_3 \in (a_2, b]$ at step 2, the strategies described above are used to try to
reduce $\hat{\varphi}$. Then a_3 is chosen, and perhaps reduced at step 4, as described above.

The reader who is not very interested in the murky details of procedure
glomin, or in the effect of rounding errors, would be well advised to skip
the rest of this section.

Some of the formulas used by procedure *glomin* need an explanation.
When either the random or nonrandom search strategy is performed, we
have numbers q and r, and wish to determine if the relation

$$q \neq 0 \wedge \left(a_2 < a_2 + \frac{r}{q} < b \right)$$

$$\wedge \frac{\left(b - \left(a_2 + \frac{r}{q} \right) \right) f(a_2) + \frac{r}{q} f(b)}{b - a_2} - \frac{Mr}{2q} \left(b - \left(a_2 + \frac{r}{q} \right) \right) < \hat{\varphi} - t$$

$$(3.13)$$

is true. If $m_2 = \frac{1}{2}M > 0$, $z_2 = b - a_2 > 0$, $y_b = f(b)$, and $y_2 = f(a_2)$, then
(3.13) is equivalent to

$$q[r(y_b - y_2) + z_2 q(y_2 - \hat{\varphi} + t)] < z_2 m_2 r(z_2 q - r), \qquad (3.14)$$

which is the condition tested after label "retry" of procedure *glomin*. (If
$q = 0$ then (3.14) is false, and it is also false if $a_2 + r/q$ lies outside (a_2, b),
since $m_2 > 0$ and $\hat{\varphi} - t < \min(y_2, y_b)$.)

To approximate a_3^*, we need the point a_3^{**} where the parabola $y = P(x)$,
passing through (a_i, y_i) for $i = 0, 1, 2$, intersects the parabola

$$y = m_2 \left(x - a_2 - \sqrt{\frac{y_2 - \hat{\varphi} + t}{m_2}} \right)^2 + \hat{\varphi} - t. \qquad (3.15)$$

(In procedure *glomin* we use c in place of a_1 to save a storage location.) Let

$z_0 = y_2 - y_1, z_1 = y_2 - y_0, d_0 = a_2 - a_1, d_1 = a_2 - a_0$, and $d_2 = a_1 - a_0$. In the nonrandom search we have already computed numbers p and q_s (r and q above) with

$$p = d_1^2 z_0 - d_0^2 z_1 \tag{3.16}$$

and

$$q_s = 2(d_0 z_1 - d_1 z_0), \tag{3.17}$$

in order to find the turning point $a_2 + p/q_s$ of $P(x)$. By forming the quadratic equation for a_3^{**}, and dividing out the unwanted root a_2, we find that

$$a_3^{**} = a_2 + \frac{p'}{q'}, \tag{3.18}$$

where

$$p' = p + 2rs, \tag{3.19}$$

$$q' = r + \tfrac{1}{2} q_s, \tag{3.20}$$

$$r = d_0 d_1 d_2 m_2, \tag{3.21}$$

and

$$s = \sqrt{\frac{y_2 - \hat{\phi} + t}{m_2}}. \tag{3.22}$$

Finally, there is the inspection of the lower bound on f in (a_2, a_3) given by the parabola

$$y = \frac{(a_3 - x)y_2 + (x - a_2)y_3}{d_0} - m_2(x - a_2)(a_3 - x), \tag{3.23}$$

where $m_2 = \tfrac{1}{2} M > 0$ and

$$d_0 = a_3 - a_2 > 0. \tag{3.24}$$

If

$$p = \frac{y_2 - y_3}{m_2 d_0}, \tag{3.25}$$

then the parabola (3.23) is monotonic increasing or decreasing in (a_2, a_3) provided

$$|p| \geq d_0. \tag{3.26}$$

Otherwise, the parabola (3.23) attains its minimum in (a_2, a_3), and the minimum value is $\tfrac{1}{2}(y_2 + y_3) - \tfrac{1}{4} m_2(d_0^2 + p^2)$ at $x = \tfrac{1}{2}(a_2 + a_3 + p)$. Thus, at step 4 of the basic algorithm, a_3 must be reduced if

$$|p| < d_0 \,\wedge\, \tfrac{1}{2}(y_2 + y_3) - \tfrac{1}{4} m_2(d_0^2 + p^2) < \hat{\phi} - t, \tag{3.27}$$

i.e., if

$$|p| < d_0 \,\wedge\, \tfrac{1}{4} M(d_0^2 + p^2) > (y_2 - \hat{\phi}) + (y_3 - \hat{\phi}) + 2t. \tag{3.28}$$

The effect of rounding errors

So far we have ignored the effect of rounding errors, which actually occur both in the computation of $f(x)$ and in the internal computations of procedure *glomin*. Now we show how these rounding errors can be accounted for.

Let ϵ be the relative machine precision (parameter *macheps* of procedure *glomin*), i.e.,

$$\epsilon = \begin{cases} \beta^{1-\tau} & \text{(truncated arithmetic)}, \\ \frac{1}{2}\beta^{1-\tau} & \text{(rounded arithmetic)}, \end{cases}$$

for τ-digit floating-point arithmetic to base β. We suppose, following Wilkinson (1963), that

$$fl(x \# y) = (x \# y)(1 + \delta), \tag{3.29}$$

where $\#$ stands for any of the arithmetic operations $+, -, \times, /$, and

$$|\delta| \leq \epsilon. \tag{3.30}$$

On machines without guard digits, the relations (3.29) and (3.30) may fail to hold for addition and subtraction: we may only have the weaker relation

$$\left. \begin{aligned} fl(x \pm y) &= x(1 + \delta_1) \pm y(1 + \delta_2), \\ |\delta_i| &\leq \epsilon \quad \text{for} \quad i = 1, 2. \end{aligned} \right\} \tag{3.31}$$

where

With these machines it seems difficult to be sure that rounding errors committed inside procedure *glomin* are harmless. At any rate, our analysis depends heavily on relation (3.29). (See equation (3.52) and the following analysis.)

We also suppose that square roots are computed with a small relative error, say

$$\left. \begin{aligned} fl(\text{sqrt}(x)) &= (1 + 3\delta)\sqrt{x}, \\ |\delta| &\leq \epsilon. \end{aligned} \right\} \tag{3.32}$$

where

(Any good square root routine should satisfy (3.32) very easily. The library routines for IBM 360 computers certainly do: see Clark, Cody, Hillstrom, and Thieleker (1967).)

Let us first consider the effect of rounding errors in the computation of f, supposing for the moment that the internal computations of procedure *glomin* are done exactly. The user has to provide procedure *glomin* with a positive tolerance e which gives a bound on the absolute error in computing f. More precisely, we assume that, for all δ and x with $|\delta| \leq \epsilon$ and $x, x(1 + \delta)$ in $[a, b]$, we have

$$|fl(f(x(1 + \delta))) - f(x)| \leq e, \tag{3.33}$$

where $f(x)$ is the exact mathematical function (satisfying condition (2.1)), and $fl(f(x))$ is its computed floating-point approximation. The reason for condition (3.33) will be apparent later: at present we only need the special case with $\delta = 0$, i.e.,

$$|fl(f(x)) - f(x)| \le e \tag{3.34}$$

for all $x \in [a, b]$.

We have seen that, without rounding errors, procedure *glomin* would return $\hat{\varphi}$ (or $y = glomin$) and $\hat{\mu}$ (or x) satisfying

$$\varphi_f \le \hat{\varphi} = f(\hat{\mu}) \le \varphi_f + t. \tag{3.35}$$

With rounding errors, (3.35) no longer holds, but we shall show that

$$\varphi_f \le f(\hat{\mu}) \le \varphi_f + t + 2e \tag{3.36}$$

and

$$\varphi_f - e \le \hat{\varphi} = fl(f(\hat{\mu})) \le \varphi_f + t + e. \tag{3.37}$$

If the error e in computing f is much less than the tolerance t, then (3.36) and (3.37) are much the same as (3.35), so rounding errors have little effect on the accuracy of $\hat{\varphi}$.

The left hand inequality in (3.36) is obvious from the definition of φ_f. To prove the right hand inequality, we must look closely at the "critical" sections of procedure *glomin*, i.e., the sections where rounding errors could make an essential difference. (Examples of noncritical sections are the random and nonrandom searches.)

In computing the safe choice a'' for a_3 according to equation (3.9), we compute

$$s = \sqrt{\frac{y_2 - \hat{\varphi} + t}{m_2}} \tag{3.38}$$

and

$$r = -\frac{1}{2}\left(d_0 + \frac{(z_0 + 2.01e)}{d_0 m_2}\right), \tag{3.39}$$

where $d_0 = a_2 - a_1$, $z_0 = y_2 - y_1$, $m_2 = \frac{1}{2}M$, $\hat{\varphi} = fl(f(\hat{\mu}))$, and $y_i = fl(f(a_i))$ for $i = 1, 2$. Thus

$$s \le \sqrt{\frac{f(a_2) - f(\hat{\mu}) + (t + 2e)}{m_2}}, \tag{3.40}$$

so, as far as the computation of s is concerned, everything said above holds if t is replaced by $t + 2e$. (Remember that we are regarding all computations inside the procedure as exact.) We are only interested in r when $d_0 > 0$ and $m_2 > 0$, and as

$$z_0 + 2.01e > z_0 + 2e \ge f(a_2) - f(a_1),$$

we have

$$r \le -\frac{1}{2}\left(d_0 + \frac{f(a_2) - f(a_1)}{d_0 m_2}\right). \tag{3.41}$$

(The reason for the extra $0.01e$ will be apparent later.) Thus, the computed a_3'' will not exceed the correct value, given by (3.9), if t is replaced by $t + 2e$.

The other point where rounding errors in the computation of f are critical is when we determine whether the parabola $y = P(x)$, with $P''(x) = M$, $P(a_2) = y_2$, and $P(a_3) = y_3$, lies above the line $y = \hat{\varphi} - t$ in the interval (a_2, a_3). Let $y = Q(x)$ be the parabola with $Q''(x) = M$, $Q(a_2) = f(a_2)$, and $Q(a_3) = f(a_3)$. Since

$$y_i = fl(f(a_i)) \leq f(a_i) + e \quad \text{for} \quad i = 2, 3,$$

it is clear that

$$P(x) \leq Q(x) + e \qquad (3.42)$$

in (a_2, a_3). Thus, if

$$P(x) \geq \hat{\varphi} - t \qquad (3.43)$$

in (a_2, a_3), then

$$Q(x) \geq \hat{\varphi} - t - e \geq f(\hat{\mu}) - t - 2e \qquad (3.44)$$

in (a_2, a_3), so again everything is accounted for by changing t to $t + 2e$. This completes the proof of (3.36). The left inequality in (3.37) is obvious, and the right inequality follows from the above argument if we note that it is sufficient to replace t by $t + e + (f(\hat{\mu}) - \hat{\varphi})$.

Now, let us consider the effect of rounding errors committed inside procedure *glomin*. We shall show that (3.36) and (3.37) still hold, provided some minor modifications are made in the algorithm. These modifications are included in procedure *glomin*, but, to avoid confusion, they were not mentioned in the description above. The most important modification is that, instead of having $m_2 = \frac{1}{2}M$, procedure *glomin* has

$$m_2 = fl(\tfrac{1}{2}(1 + 16\epsilon)M), \qquad (3.45)$$

where the factor $1 + 16\epsilon$ is introduced purely to nullify the effect of rounding errors.

For the sake of simplicity, terms of order ϵ^2 are ignored in the rest of this section. Because of the slack in some of our inequalities, these terms may be accounted for if $\epsilon \leq \frac{1}{400}$. From (3.45) and the assumption (3.29), we certainly have

$$m_2 \geq \tfrac{1}{2}(1 + 13\epsilon)M. \qquad (3.46)$$

In the computation of a_3'' according to (3.9), procedure *glomin* actually computes

$$\bar{s} = fl\left(\frac{(y_2 - \hat{\varphi}) + t}{m_2}\right)^{1/2}, \qquad (3.47)$$

and since errors in the computation of f have already been accounted for, we can assume that y_2 and $\hat{\varphi}$ are exact floating-point numbers. From (3.46) and the assumptions (3.29) and (3.32),

$$\bar{s} \leq (1 + 3\delta_4)\left(\frac{[(y_2 - \hat{\phi})(1 + \delta_1) + t](1 + \delta_2)(1 + \delta_3)}{\frac{1}{2}M(1 + 13\epsilon)}\right)^{1/2}, \qquad (3.48)$$

where $|\delta_i| \leq \epsilon$ for $i = 1, \ldots, 4$. Since $y_2 - \hat{\phi}$ and t are both nonnegative,

$$(y_2 - \hat{\phi})(1 + \epsilon) + t \leq (y_2 - \hat{\phi} + t)(1 + \epsilon), \qquad (3.49)$$

so

$$\bar{s} \leq s = \left(\frac{y_2 - \hat{\phi} + t}{\frac{1}{2}M}\right)^{1/2}. \qquad (3.50)$$

Thus, the slight modification of m_2 has ensured that the computed s is no greater than the exact s. Note that, in the derivation of (3.50), it is essential that $y_2 - \hat{\phi}$ is computed with a small relative error, so the assumption (3.29) is necessary: (3.31) is not enough.

Similarly, to find a_3'', we actually compute

$$\bar{r} = fl\left[-\frac{1}{2}\left((a_2 - a_1) + \frac{(y_2 - y_1) + 2.01e}{(a_2 - a_1)m_2}\right)\right], \qquad (3.51)$$

where $e > 0$, $m_2 > 0$, and $a_2 > a_1$. We are only interested in $\bar{r}$ if $\bar{r} > 0$, so

$$\begin{aligned}
0 &> fl((y_2 - y_1) + 2.01e) \\
&\geq ((y_2 - y_1)(1 + \epsilon) + 2.01e(1 - \epsilon))(1 + \epsilon) \\
&\geq (y_2 - y_1 + 2e)(1 + \epsilon)^2, \qquad (3.52)
\end{aligned}$$

assuming that $\epsilon \leq \frac{1}{400}$. (The reason for the extra $0.01e$ in (3.39) is now clear.) Thus

$$\bar{r} = fl(-\frac{1}{2}(r_1 + r_2)), \qquad (3.53)$$

where

$$0 < (a_2 - a_1)(1 - \epsilon) \leq r_1 \leq (a_2 - a_1)(1 + \epsilon) \qquad (3.54)$$

and

$$0 > r_2 \geq \frac{(y_2 - y_1 + 2e)(1 - 9\epsilon)}{\frac{1}{2}M(a_2 - a_1)}. \qquad (3.55)$$

Since $\bar{r} > 0$, (3.53) shows that $|r_1| < |r_2|$, so, from (3.53) to (3.55),

$$\bar{r} \leq r \leq -\frac{1}{2}\left[(a_2 - a_1) + \left(\frac{y_2 - y_1 + 2e}{\frac{1}{2}M(a_2 - a_1)}\right)\right]. \qquad (3.56)$$

As before, the computed $\bar{r}$ is no greater than the correct r. The same is not true for $\bar{a}_3''$, the computed value of a_3'', but $\bar{a}_3''$ is either b, $fl(a_2 + \bar{r})$, or $fl(a_2 + \bar{s})$. Suppose, for example, that

$$\bar{a}_3'' = fl(a_2 + \bar{s}). \qquad (3.57)$$

Then

$$fl(f(\bar{a}_3'')) = fl\{f[(a_2 + \bar{s})(1 + \delta)]\} \qquad (3.58)$$

where $|\delta| \leq \epsilon$, so, from (3.33),

$$|fl[f(\tilde{a}_3'')] - f(a_2 + \tilde{s})| \leq e. \qquad (3.59)$$

(This is why we required (3.33) instead of the weaker (3.34).) Thus, the error in computing $a_2 + \tilde{s}$ or $a_2 + \tilde{r}$ can be ignored, for it has been absorbed into the assumption (3.33) on e.

Finally, we have to consider the effect of rounding errors when testing the condition (3.28). First

$$\tilde{p} = fl\left(\frac{y_2 - y_3}{\frac{1}{2}M(a_3 - a_2)}\right) \qquad (3.60)$$

is computed. It is important to note that we use $\frac{1}{2}M$, not the slightly different m_2 (given by (3.45)) here. Thus

$$\tilde{p} = \left(\frac{y_2 - y_3}{\frac{1}{2}M(a_3 - a_2)}\right)(1 + 5\delta_1) \qquad (3.61)$$

and

$$\tilde{d}_0 = fl(a_3 - a_2) = (a_3 - a_2)(1 + \delta_2), \qquad (3.62)$$

where $|\delta_i| \leq \epsilon$ for $i = 1, 2$.

The test actually made by procedure *glomin* is whether

$$|\tilde{p}| < fl((1 + 9\epsilon)\tilde{d}_0) \wedge fl(\frac{1}{2}m_2(\tilde{d}_0^2 + \tilde{p}^2)) > fl[(y_2 - \hat{\phi}) + (y_3 - \hat{\phi}) + 2t], \qquad (3.63)$$

and we shall show that (3.63) is true whenever the condition (3.28) is true. First, $|p| < d_0$ implies that $|\tilde{p}| < d_0(1 + 5\epsilon)$, and thus

$$|\tilde{p}| < fl((1 + 9\epsilon)\tilde{d}_0). \qquad (3.64)$$

Similarly, if $|p| < d_0$ and

$$\frac{1}{4}M(d_0^2 + p^2) > (y_2 - \hat{\phi}) + (y_3 - \hat{\phi}) + 2t, \qquad (3.65)$$

then

$$\tilde{d}_0^2 + \tilde{p}^2 \geq (d_0^2 + p^2)(1 - 6\epsilon), \qquad (3.66)$$

so

$$fl(\frac{1}{2}m_2(\tilde{d}_0^2 + \tilde{p}^2)) \geq \frac{1}{4}M(d_0^2 + p^2)(1 + 4\epsilon)$$
$$> [(y_2 - \hat{\phi}) + (y_3 - \hat{\phi}) + 2t](1 + 3\epsilon)$$
$$\geq fl[(y_2 - \hat{\phi}) + (y_3 - \hat{\phi}) + 2t]. \qquad (3.67)$$

(Note the importance of grouping the terms: since $y_2 - \hat{\phi}$, $y_3 - \hat{\phi}$, and $2t$ are all nonnegative, their sum can be computed with a small relative error.)

From (3.64) and (3.67), the inexact test (3.63) results in a_3 being reduced whenever the exact test (3.28) says that it must be. a_3 may occasionally be reduced unnecessarily because of rounding errors, but this does not invalidate the bounds (3.36) and (3.37); it merely causes some unnecessary function evaluations.

We should mention a remote possibility that rounding errors can prevent convergence. This is only possible if $fl(a_2 + \tilde{s}) = a_2$ and, as

$\bar{s} \geq (1 - 14\epsilon)(2t/M)^{1/2}$, it is impossible if

$$t \geq M\epsilon^2 \max(a^2, b^2). \tag{3.68}$$

Thus, convergence can only be prevented by rounding errors if t is unreasonably small.

In conclusion, procedure *glomin* is guaranteed to return $\hat{\varphi}$ and $\hat{\mu}$ satisfying the bounds (3.36) and (3.37), provided the input parameters *macheps*, t, and e are set correctly.

Section 4
THE RATE OF CONVERGENCE IN SOME SPECIAL CASES

It is difficult to say much in general about the number of function evaluations required by the algorithm described in Section 3. In the next section we compare the algorithm with the best possible one for given M and t. In this section, we try to gain some insight into the dependence of the number of function evaluations on the bound M and the tolerance t, by looking at some simple special cases.

The worst case

As pointed out above (equation (3.4)), two function evaluations are enough to determine $\hat{\mu}$ and $\hat{\varphi}$ if $M \leq 0$, so suppose that $M > 0$, and let

$$\delta = \sqrt{\frac{2t}{M}}. \tag{4.1}$$

We showed above that, if the last function evaluation was at $a_2 \in [a, b]$, we could safely choose

$$a_3 = \min(b, a_2 + \delta) \tag{4.2}$$

for the next evaluation (step 2 of the basic algorithm). With this simple choice of a_3, about $(b - a)/\delta$ function evaluations would be required. Procedure *glomin* tries to do better than this, and is nearly always successful (see Section 6), but the worst that can happen is that a_3 will be chosen to be b, and then a_3 will be reduced several times at step 4 of the basic algorithm. As $a_3 - a_2$ is halved at each such reduction of a_3, there can be at most

$$\left\lceil \log_2\left(\frac{b - a_2}{\delta}\right)\right\rceil \leq \left\lceil \log_2\left(\frac{b - a}{\delta}\right)\right\rceil \tag{4.3}$$

consecutive reductions of a_3 at step 4. Thus, at worst, about

$$\left(\frac{b - a}{\delta}\right) \log_2\left(\frac{b - a}{\delta}\right) \tag{4.4}$$

function evaluations will be required. We have ignored the random and nonrandom searches, but these can only add about $2(b - a)/\delta$ extra function evaluations.

If δ is given by (4.1), the term $\log_2 (b - a)/\delta$ in (4.4) varies only slowly with M and t, so the upper bound is roughly proportional to $(b - a)(M/t)^{1/2}$. In particular, the upper bound is roughly proportional to $\sqrt{M}$, and it seems to be a good general rule that the number of function evaluations is roughly proportional to $\sqrt{M}$, even when the upper bound (4.4) is not attained (see also Section 6).

A straight line

If the global minimum of f occurs at an endpoint $\mu = a$ or b, and $f'(\mu) \neq 0$, we can obtain insight into the behavior of the algorithm near μ by considering the linear approximation $f(\mu) + (x - \mu)f'(\mu)$ to $f(x)$. Suppose, for example, that

$$f(x) = k(x - a) + t \tag{4.5}$$

for some $k > 0$, so $\mu = a$. Ignoring the random searches, the algorithm will evaluate f at the points a, b, c, and then at points $x_1 < x_2 < x_3 < \cdots < x_{N-1}$. Here $x_0 = a < x_1$, $x_N \geq b$, and the points $(x_n, f(x_n))$ and $(x_{n+1}, f(x_{n+1}))$ lie on the parabola $y = P_n(x)$ which touches the line $y = 0$ and has $P_n''(x) = M$. (See Diagram 4.1.) If $P_n(x)$ touches $y = 0$ at $x = \alpha_n$, then

$$P_n(x) = \tfrac{1}{2}M(x - \alpha_n)^2, \tag{4.6}$$

so

$$\alpha_n = x_n + \sqrt{\frac{2}{M}(k(x_n - a) + t)} = x_{n+1} - \sqrt{\frac{2}{M}(k(x_{n+1} - a) + t)}. \tag{4.7}$$

If

$$z_n = \sqrt{x_n - a + \frac{t}{k}}, \tag{4.8}$$

then (4.7) gives

$$z_{n+1} = z_n + \sqrt{\frac{2k}{M}}, \tag{4.9}$$

so

$$z_n = \sqrt{\frac{t}{k}} + n\sqrt{\frac{2k}{M}}. \tag{4.10}$$

Thus

$$x_n = a + n\sqrt{\frac{8t}{M}} + n^2\left(\frac{2k}{M}\right), \tag{4.11}$$

and as N is the least positive n such that $x_n \geq b$, this gives

$$N = \left\lceil \frac{\sqrt{\frac{1}{2}M}}{k} \left(\sqrt{k(b-a)+t} - \sqrt{t} \right) \right\rceil. \tag{4.12}$$

(4.12) shows that N is essentially proportional to $\sqrt{M}$.

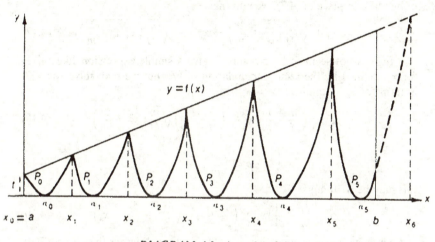

DIAGRAM 4.1 A straight line

Two limiting cases of (4.12) are interesting. If t is small and k not too small, so that $k(b-a) \gg t$, then

$$N \simeq \sqrt{\frac{M(b-a)}{2k}}, \tag{4.13}$$

which is independent of t. (In this section we are neglecting the effect of rounding errors, but these should not be important if t satisfies the weak condition (3.68).)

If k is very small, so that $k(b-a) \ll t$, then (4.12) gives

$$N \simeq \frac{b-a}{2\delta}, \tag{4.14}$$

and the algorithm proceeds in steps of size about 2δ, where δ is given by (4.1).

A parabola

If the global minimum of f occurs at an interior point μ, then $f'(\mu) = 0$. If $f''(\mu) \neq 0$ we may analyze the behavior of the algorithm near μ by considering the parabolic approximation $f(\mu) + \frac{1}{2}f''(\mu)(x-\mu)^2$ to $f(x)$. Thus, suppose that

$$M > m > 0 \tag{4.15}$$

and

$$f(x) = \tfrac{1}{2}m(x-\mu)^2 + t, \tag{4.16}$$

where $\mu \in (a, b)$. The nonrandom search will quickly locate μ, so we may suppose that $\hat{\mu} = \mu$ and, without loss of generality, $\mu = 0$. The algorithm will call for the evaluation of f at points to the left, and then to the right, of μ. As these two cases are similar, let us define $x_0 = \mu = 0$, and study the points $x_1, x_2, \ldots$ defined above, except that now f is given by (4.16) instead of by (4.5). In place of (4.7), we find that

$$\alpha_n = x_n + \sqrt{\frac{m}{M}\left(x_n^2 + \frac{2t}{m}\right)} = x_{n+1} - \sqrt{\frac{m}{M}\left(x_{n+1}^2 + \frac{2t}{m}\right)}. \quad (4.17)$$

It does not seem to be possible to give a simple expression like (4.11) for x_n, defined by the recurrence relation (4.17), but we may solve for x_{n+1} in terms of x_n, obtaining

$$x_{n+1} = \left(\frac{M+m}{M-m}\right)x_n + \left(\frac{2M}{M-m}\right)\sqrt{\frac{m}{M}\left(x_n^2 + \frac{2t}{m}\right)}. \quad (4.18)$$

If

$$\rho = \left(\frac{M}{m}\right)^{1/2}, \quad (4.19)$$

this may be written as

$$x_{n+1} = \left(\frac{\rho+1}{\rho-1}\right)x_n + \left(\frac{2\rho}{\rho^2-1}\right)\left(\sqrt{x_n^2 + \frac{2t}{m}} - x_n\right). \quad (4.20)$$

Suppose that ρ is close to 1, i.e., M is not much larger than $m = f''(\mu)$. Then

$$x_1 = \left(\frac{2\rho}{\rho^2-1}\right)\sqrt{\frac{2t}{m}} \quad (4.21)$$

and, for $n \geq 1$,

$$x_{n+1} = \left(\frac{\rho+1}{\rho-1}\right)x_n\{1 + O[(\rho-1)^2]\} \quad \text{as} \quad \rho \to 1. \quad (4.22)$$

Thus

$$x_n \simeq \left(\frac{\rho+1}{\rho-1}\right)^n \sqrt{\frac{t}{2m}}. \quad (4.23)$$

As the factor $(\rho + 1)/(\rho - 1)$ is large, only a few function evaluations will be required.

Section 5
A LOWER BOUND ON THE NUMBER OF FUNCTION EVALUATIONS REQUIRED

Suppose that a positive tolerance t and bound M are given, that f attains its global minimum φ_f in $[a, b]$ at μ_f, and that

$$f''(x) < M \quad (5.1)$$

for all $x \in [a, b]$. (Similar results to those below hold if equality is allowed, but the definitions and proofs have to be modified slightly.) First, we need a lemma.

LEMMA 5.1

If $x' \in [a, b)$, then there is at most one point $x'' \in (x', b]$ such that the parabola $y = P(x)$, with $P''(x) = M$, $P(x') = f(x')$, and touching the line $y = \varphi_f - t$, satisfies $P(x'') = f(x'')$.

Proof

Suppose, by way of contradiction, that two such distinct points x'' and x''' exist. Then

$$M = 2f[x', x'', x'''] = f''(\xi) \tag{5.2}$$

for some $\xi \in [x', b]$ (see Chapter 2), contradicting

$$f''(\xi) < M. \tag{5.3}$$

DEFINITION 5.1

For $x' \in [a, b)$, define

$$s(x') = \begin{cases} x'' & \text{if the point } x'' \text{ of Lemma 5.1 exists,} \\ b & \text{otherwise.} \end{cases}$$

LEMMA 5.2

If $x \in [a, b)$ and $s(x) \neq b$, then

$$s(x) - x \geq \sqrt{\frac{8t}{M}}. \tag{5.4}$$

Proof

This follows by considering the parabola, with second derivative M, which passes through $(x, f(x))$ and $(s(x), f[s(x)])$, and touches the line $y = \varphi_f - t$, since $f(x) \geq \varphi_f$ and $f[s(x)] \geq \varphi_f$.

DEFINITION 5.2

An integer N and points $a' = x_1 < x_2 < x_3 < \ldots < x_N = b$ are defined thus: $x_1 = a$ and, for $n \geq 2$ and $x_{n-1} < b$, $x_n = s(x_{n-1})$. (See Diagram 5.1.)
Lemma 5.2 shows that N is finite, in fact

$$N \leq 1 + \left\lceil (b - a)\left(\frac{M}{8t}\right)^{1/2} \right\rceil. \tag{5.5}$$

The following lemma shows that in order to prove that $f(x) \geq \varphi_f - t$ for all $x \in [a, b]$, given only condition (5.1), it is sufficient to evaluate f at $x_1, x_2, \ldots, x_N$.

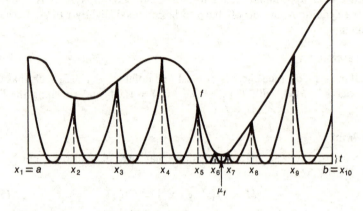

DIAGRAM 5.1 The points $x_1, \ldots, x_N$

LEMMA 5.3

If $g \in C^2[a, b]$, $g''(x) \leq M$ for all $x \in [a, b]$, and

$$g(x_n) = f(x_n) \qquad (5.6)$$

for $n = 1, 2, \ldots, N$, where the points x_n are defined above, then

$$\varphi_g \geq \varphi_f - t. \qquad (5.7)$$

The lemma follows immediately from the definitions and Theorem 2.1. Our interest in the points $x_1, \ldots, x_N$ stems from the following theorem, which complements Lemma 5.3.

THEOREM 5.1

Let $x_1' < x_2' < \ldots < x_\nu'$ be any ν points in $[a, b]$, with $\nu < N$. Then there is a function $g \in C^\infty[a, b]$, satisfying

$$g''(x) < M \qquad (5.8)$$

for all $x \in [a, b]$, and

$$g(x_n') = f(x_n') \qquad (5.9)$$

for $n = 1, 2, \ldots, \nu$, such that

$$\varphi_g < \varphi_f - t. \qquad (5.10)$$

Proof

Suppose, by way of contradiction, that

$$\varphi_g \geq \varphi_f - t \qquad (5.11)$$

for all such g. Then $x_1' = a$, for otherwise $-g(a)$ can be arbitrarily large, and, similarly, $x_\nu' = b$. Since $\nu < N$, there is an n, $1 \leq n < \nu$, such that $x_n' \leq x_n$ and $x_{n+1}' > x_{n+1}$. Thus, the parabola $y = P(x)$, with $P''(x) = M$, $P(x_n') =$

$f(x'_n)$, and $P(x'_{n+1}) = f(x'_{n+1})$, is such that

$$\min_{x \in [x_n', \, x_{n+1}']} P(x) < \varphi_f - t. \tag{5.12}$$

Since there is a function g as above which is arbitrarily close to $P(x)$ in $[x_n', x_{n+1}']$, this contradicts (5.11), so the theorem holds.

Consequences of the theorem

Theorem 5.1 says that, if all that is known *a priori* about f is that $f \in C^2[a, b]$ and satisfies condition (5.1), then any algorithm which is guaranteed to find $\hat{\mu}$ so that $f(\hat{\mu}) \leq \varphi_f + t$ must require at least N evaluations of f. If an algorithm required only $v < N$ evaluations, at points $x_1' < x_2' < \ldots < x_v'$, then it would be sure to fail for either f or for g, for f and g are indistinguishable on the basis of the v function evaluations, but $\varphi_g + t < \varphi_f$. Of course, we are only considering algorithms which sequentially evaluate f at a finite number of points.

Conversely, Lemma 5.3 implies that $N + 1$ function evaluations are sufficient: just evaluate f at μ_f and at $x_1, \ldots, x_N$. (See Diagram 5.1.) Unfortunately, Lemma 5.3 does not give us an effective algorithm for approximating φ_f, for we do not know N or the points $x_2, \ldots, x_{N-1}$ in advance.

Efficiency

Suppose that an algorithm requires N' function evaluations to find $\hat{\varphi} = f(\hat{\mu})$ such that $\hat{\varphi} \leq \varphi_f + t$ is guaranteed. We could define the *efficiency* E of the algorithm by

$$E = N/N'. \tag{5.13}$$

(Note that E depends on f, M, t, a, and b, as well as on the algorithm.) We have shown that

$$E \leq 1 \tag{5.14}$$

for any correct (i.e., guaranteed) algorithm. Thus, if an algorithm has an efficiency close to 1, we are justified in saying that the algorithm is nearly optimal for that f, M, t, etc. In the next section we give numerical results which show that the algorithm described in Section 3 is often nearly optimal.

Section 6
PRACTICAL TESTS

The ALGOL procedure *glomin* given in Section 10 was tested using ALGOL W (Wirth and Hoare (1966); Bauer, Becker, and Graham (1968)) on an IBM 360/91 computer with machine precision 16^{-13}. Some representative numerical results are summarized in Table 6.1. For all of these results the parameters e and *macheps* were set at 10^{-14} and 16^{-13} respectively.

TABLE 6.1 **Numerical results for procedure** *glomin*

f	M	N''	N'	N	$E = N/N'$
	0	2	2	2	1.00
f_1	100	15	15	11	0.73
	10000	106	106	101	0.95
	2	4	4	2	0.50
	2.1	8	11	8	0.73
f_2	2.2	9	13	9	0.69
	8	25	34	29	0.85
	32	48	68	60	0.88
	128	95	141	120	0.85
	14	38	51	37	0.73
f_3	28	48	68	54	0.79
	56	67	98	76	0.78
f_4	72	222	246	126	0.51
f_5	72	456	542	437	0.81

The symbols are explained below. The functions are:
$f_1(x) = 2 - x$ on $[7, 9]$ (in all cases $\hat{\mu} = 9$, $\hat{\varphi} = 7$),
$f_2(x) = x^2$ on $[-1, 2]$ (in all cases $\hat{\mu} = \hat{\varphi} = 0$),
$f_3(x) = x^2 + x^3$ on $[-\frac{1}{2}, 2]$ (for $t = 10^{-12}, |\hat{\mu}| < 3 \times 10^{-10}, |\hat{\varphi}| < 6 \times 10^{-20}$),
$f_4(x) = (x + \sin x)\exp(-x^2)$ on $[-10, 10]$
$\qquad\qquad\qquad (\hat{\mu} = -0.6795786599525, \hat{\varphi} = -0.824239398476077)$, and
$f_5(x) = (x - \sin x)\exp(-x^2)$ on $[-10, 10]$
$\qquad\qquad\qquad (\hat{\mu} = -1.195136641665, \hat{\varphi} = -0.0634905289364399)$.

The table gives the upper bound M (parameter m of *glomin*) on f'', and the total number of function evaluations required by procedure *glomin*: N'' with tolerance $t = 10^{-8}$, and N' with tolerance $t = 10^{-12}$. The lower bound N defined in Section 5 is also given for $t = 10^{-12}$. (Recall that no algorithm which is guaranteed to succeed can take fewer than N function evaluations.) N and the points $x_1, \ldots, x_N$ of Section 5 were computed in the obvious way from Definition 5.2, using procedure *zero* of Chapter 4 to solve the nonlinear equation

$$P(x) = f(x), \qquad\qquad (6.1)$$

where $P(x)$ is the parabola of Lemma 5.1. Finally, the efficiency $E = N/N'$ (equation (5.13)) is given.

For some more numerical results, see Section 8.

Comments on Table 6.1

The results for the simple functions $f_1(x) = 2 - x$ and $f_2(x) = x^2$ verify the predictions made in Section 4. For example, the values $N = 11$ and $N = 101$ for f_1 are exactly as predicted: one more than the right side of equation (4.12). N, N', and N'' are roughly proportional to $\sqrt{M}$ if $M \gg f''(\mu)$ (see also the results for f_3), but this rule breaks down if $M \simeq f''(\mu)$, as expected from equation (4.23). (See the results for f_2 with $M = 2, 2.1, 2.2$.)

It appears that the number of function evaluations does not depend strongly on t: comparing N'' with N', we see that the average number of function evaluations required is only about twenty percent more for $t = 10^{-12}$ than for $t = 10^{-8}$.

Finally, the efficiency E of the algorithm is fairly high, even for the difficult functions f_4 and f_5. This means that no correct algorithm based entirely on function evaluations could do very much better than ours, at least on these examples.

Section 7
SOME EXTENSIONS AND GENERALIZATIONS

So far we have assumed that $f \in C^2[a, b]$ and

$$f''(x) \leq M \tag{7.1}$$

for all $x \in [a, b]$, or at least that $f \in C^1[a, b]$ and

$$f'(x) - f'(y) \leq M(x - y) \tag{7.2}$$

for $a \leq y < x \leq b$. Condition (7.2) was necessary to prove the basic Theorem 2.1. For the application discussed in Section 8 (global minimization of a function of several variables), we need to find the global minimum of a function which is continuous, but not necessarily differentiable. We can justify using procedure *glomin*, even though f may not be differentiable, because of the following Theorems 7.1 to 7.3, which generalize Theorems 2.1 to 2.3. If the reader is prepared to accept the fact that Theorems 2.1 to 2.3 can be generalized in the appropriate way, he may skip this section.

THEOREM 7.1

Let $f \in C[a, b]$, and suppose that there is a constant M such that, for all sufficiently small $h > 0$,

$$f(u + h) - 2f(u) + f(u - h) \leq Mh^2 \tag{7.3}$$

for all $u \in [a + h, b - h]$. Then, for all $x \in [a, b]$,

$$f(x) \geq \frac{(b - x)f(a) + (x - a)f(b)}{b - a} - \frac{1}{2}M(x - a)(b - x). \tag{7.4}$$

Proof

There is no loss of generality in assuming that

$$f(a) = f(b) = 0 \tag{7.5}$$

and

$$M = 0, \tag{7.6}$$

for we can consider $f(x) - P(x)$, where $P(x)$ is the right side of (7.4), instead of $f(x)$. Thus, we have to show that

$$\varphi_f \geq 0, \tag{7.7}$$

where φ_f is the least value of f on $[a, b]$. Suppose, by way of contradiction, that

$$\varphi_f < 0, \tag{7.8}$$

and let

$$u = \sup\{x \in [a, b] \,|\, f(x) = \varphi_f\}. \tag{7.9}$$

By the continuity of f, $f(u) = \varphi_f < 0$, so $u \neq a$ or b. Thus, for sufficiently small $h > 0$, $u \in [a + h, b - h]$ and, from the definition of u,

$$f(u - h) \geq f(u) \tag{7.10}$$

and

$$f(u + h) > f(u). \tag{7.11}$$

Because of the assumption (7.6), this contradicts (7.3), so (7.8) is impossible, and the result follows. (Note the close connection with the maximum principle for elliptic difference operators.)

THEOREM 7.2

Suppose that (7.3) holds, $M > 0$, $a \leq c_1 < c_2 \leq b$, and $f(a) \geq f(c_1) = f(c_2)$. Then

$$c_2 - a > \sqrt{\frac{f(a) - f(c_1)}{\frac{1}{2}M}}. \tag{7.12}$$

Proof

Apply Theorem 7.1 with x replaced by c_1 and b by c_2. The hypothesis that $f(c_1) = f(c_2)$ gives, after some simplification,

$$(c_1 - a)(c_2 - a) \geq \frac{f(a) - f(c_1)}{\frac{1}{2}M}, \tag{7.13}$$

and the result follows since $c_2 - a > c_1 - a \geq 0$.

THEOREM 7.3

Suppose that (7.3) holds, $M > 0$, $a < c \leq b$, and the interval $I = [c, b] \cap [c, \frac{1}{2}(a + c) - \{f(a) - f(c)\}/\{M(a - c)\}]$ has positive length. Then $f(x)$ is strictly monotonic decreasing on I.

Proof

Suppose $x_1, x_2 \in I$ with $x_1 < x_2$. We have to show that

$$f(x_1) > f(x_2). \tag{7.14}$$

Apply Theorem 7.1, first with x replaced by c and b by x_1, then with a replaced by c, x by x_1 and b by x_2. The two resulting inequalities give, after some simplification,

$$\frac{f(x_1) - f(x_2)}{M(x_2 - x_1)} \geq \frac{a + c}{2} - \frac{f(a) - f(c)}{M(a - c)} - \frac{x_1 + x_2}{2}. \tag{7.15}$$

The right side of (7.15) is positive, so (7.14) holds.

Remarks

Theorems 7.1 to 7.3 generalize Theorems 2.1 to 2.3 respectively. Since the algorithm described in Section 3 is based entirely on Theorems 2.1 to 2.3, it is clear that condition (7.3) is sufficient for the algorithm to find a correct approximation to the global minimum of f. This is not surprising, for condition (7.3) is equivalent to (7.2) if $f \in C^1[a, b]$, and to (7.1) if $f \in C^2[a, b]$. In the next section, we use this result to develop an algorithm for finding the global minimum of a function f of several variables. The conditions on f are much weaker than those required by Newman (1965), Sugie (1964), or Krolak and Cooper (1963). (See also Kaupe (1964) and Kiefer (1957).)

Section 8
AN ALGORITHM FOR GLOBAL MINIMIZATION OF A FUNCTION OF SEVERAL VARIABLES

Suppose that $D = [a_x, b_x] \times [a_y, b_y]$ is a rectangle in R^2, $f: D \to R$ has continuous second derivatives on D, and constants M_x and M_y are known such that

$$f_{xx}(x, y) \leq M_x \tag{8.1}$$

and

$$f_{yy}(x, y) \leq M_y, \tag{8.2}$$

for all $(x, y) \in D$. Let us define $\varphi: [a_y, b_y] \to R$ by

$$\varphi(y) = \min_{x \in [a_x, b_x]} f(x, y). \tag{8.3}$$

Clearly $\varphi(y)$ is continuous, and

$$\min_{(x,y) \in D} f(x, y) = \min_{y \in [a_y, b_y]} \varphi(y). \tag{8.4}$$

Thus, we have reduced the minimization of $f(x, y)$, a function of two variables, to the minimization of functions of one variable. Procedure *glomin* (see Sections 3 and 10) can be used to evaluate $\varphi(y)$ for a given y, using condition

(8.1). If we could show that

$$\varphi''(y) \leq M_y, \tag{8.5}$$

then procedure *glomin* could be used again (recursively) to minimize $\varphi(y)$, and thus, from (8.4), $f(x, y)$. Unfortunately, $\varphi(y)$ need not be differentiable everywhere in $[a_y, b_y]$, so (8.5) may be meaningless. For example, consider

$$f(x, y) = xy \tag{8.6}$$

on $D = [-1, 1] \times [-1, 1]$. Then

$$\varphi(y) = \min(y, -y) = -|y|, \tag{8.7}$$

which is not differentiable at $y = 0$, and we cannot expect to prove (8.5). The same problem may arise if the minimum in (8.3) occurs at an interior point of D: one example is

$$f(x, y) = (x^3 - 3x) \sin y \tag{8.8}$$

on $D = [\sqrt{3}, \sqrt{3}] \times [-10, 10]$. ($f_x(x, y)$ vanishes for $x = \pm 1$, so $\varphi(y) = -2|\sin y|$, which is not differentiable at $0, \pm\pi$, etc.)

Fortunately, the following theorem shows that $\varphi(y)$ does satisfy a condition like (7.3), so the results of Section 7 show that procedure *glomin* can be used to find the global minimum of $\varphi(y)$ even if $\varphi(y)$ is not differentiable everywhere.

THEOREM 8.1

Let $f(x, y)$ and $\varphi(y)$ be as above. Then, for all $h > 0$ and $y \in [a_y + h, b_y - h]$,

$$\varphi(y + h) - 2\varphi(y) + \varphi(y - h) \leq M_y h^2. \tag{8.9}$$

Proof

From the definition (8.3) of $\varphi(y)$, there is a function $\mu(y)$ from $[a_y, b_y]$ into $[a_x, b_x]$ (not necessarily continuous), such that

$$\varphi(y) = f(\mu(y), y). \tag{8.10}$$

Thus

$$\varphi(y \pm h) \leq f(\mu(y), y \pm h), \tag{8.11}$$

so

$$\varphi(y + h) - 2\varphi(y) + \varphi(y - h) \leq f(\mu(y), y + h) - 2f(\mu(y), y) + f(\mu(y), y - h), \tag{8.12}$$

and the result follows from condition (8.2).

COROLLARY 8.1

For all $y \in [a_y, b_y]$ at which $\varphi''(y)$ exists,

$$\varphi''(y) \leq M_y. \tag{8.13}$$

Functions of n variables

Theorem 8.2 generalizes Theorem 8.1 to functions of any finite number of variables.

THEOREM 8.2

Suppose that $n \geq 1$; I_i is a nonempty compact set in R for $i = 1, \ldots,$ $n + 1$; $D = I_1 \times I_2 \times \cdots \times I_{n+1} \subseteq R^{n+1}$; $f\colon D \to R$ is continuous, and

$$f(\mathbf{x} + h\mathbf{e}_i) - 2f(\mathbf{x}) + f(\mathbf{x} - h\mathbf{e}_i) \leq M_i h^2 \qquad (8.14)$$

for all sufficiently small $h > 0$, all $\mathbf{x} \in R^{n+1}$ such that $\mathbf{x}, \mathbf{x} \pm h\mathbf{e}_i \in D$, and $i = 1, 2, \ldots, n + 1$. Let $D' = I_1 \times \cdots \times I_n$, and define $\varphi\colon D' \to R$ by

$$\varphi(\mathbf{y}) = \min_{x \in I_{n+1}} f(y_1, \ldots, y_n, x). \qquad (8.15)$$

Then φ is continuous on D',

$$\min_{\mathbf{x} \in D} f(\mathbf{x}) = \min_{\mathbf{y} \in D'} \varphi(\mathbf{y}), \qquad (8.16)$$

and

$$\varphi(\mathbf{y} + h\mathbf{e}'_j) - 2\varphi(\mathbf{y}) + \varphi(y - h\mathbf{e}'_j) \leq M_j h^2 \qquad (8.17)$$

for all sufficiently small $h > 0$, $y \in R^n$ such that $\mathbf{y}, \mathbf{y} \pm h\mathbf{e}'_j \in D'$, and $j = 1, 2, \ldots, n$. (Here $\mathbf{e}_i$ is a unit vector in R^{n+1}, and $\mathbf{e}'_j$ is a unit vector in R^n.)

The proof is an easy generalization of the proof of Theorem 8.1, so the details are omitted.

Theorem 8.2 shows that it is possible to use procedure *glomin* to find the global minimum of a function $f(x_1, \ldots, x_n)$ of any finite number $n \geq 1$ of variables, provided upper bounds are known for the partial derivatives $f_{x_i x_i}(\mathbf{x})$ $(i = 1, \ldots, n)$. It is interesting that no bounds on the cross derivatives $f_{x_i x_j}(\mathbf{x})$ $(i \neq j)$ are necessary.

If a one-dimensional minimization using procedure *glomin* requires about K function evaluations, then we would expect that about K^n function evaluations would be required for an n-dimensional minimization. Since K is likely to be in the range $10 \leq K \leq 100$ in practice (see Section 6), the computation involved is likely to be excessive for $n > 3$. Thus, for functions of more than three variables, we probably must be satisfied with methods which find local, but not necessarily global, minima (see Chapter 7). The theorems of Section 5 have not been extended to functions of more than one variable, so we do not know how far our procedure is from the best possible (given only upper bounds on $f_{x_i x_i}$ for $i = 1, \ldots, n$). Thus, there is a chance that a much better method for finding the global minimum of a function of several variables exists. It is also possible that slightly stronger *a priori* conditions on f (e.g., both upper and lower bounds on certain derivatives) might enable us to find the global minimum much more efficiently.

Minimization of a function of two variables: procedure glomin2d

In Section 10 we give an ALGOL 60 procedure (*glomin2d*) for finding the global minimum of a function $f(x, y)$ of two variables, using the method suggested above. Note that *glomin2d* uses procedure *glomin* in a recursive manner, for *glomin* is required both to evaluate and to minimize φ. The error bounds given in the initial comment of procedure *glomin2d* are easily derived from the error bounds (3.36) and (3.37) for procedure *glomin*.

Procedure *glomin2d* was tested on an IBM 360/91 computer (using ALGOL W), and some numerical results are summarized in Table 8.1. In all cases shown in the table the parameters *macheps*, e; and t were set at 16^{-13}, 10^{-14}, and 10^{-10} respectively. (Thus $\varphi_f - 10^{-14} \le \hat{\varphi} \le \varphi_f + 1.0002 \times 10^{-10}$ is guaranteed, where φ_f is the true minimum of f, and $\hat{\varphi}$ is the value returned by the procedure.) In the table we give the upper bounds M_x and M_y (see equations (8.1) and (8.2)), the total number of function evaluations N, and the approximate global minimum $\hat{\varphi}$ (always very close to the true global minimum φ_f).

TABLE 8.1   Numerical results for procedure *glomin2d*

f	M_x	M_y	N	$\hat{\varphi}$
f_1	0	0	4	-1
	4	4	9	-1
f_2	2	4	51	0
	2	10	116	0
	10	4	446	3'–35
	10	10	956	4'–39
f_3	2210	200	13320	2'–18
f_4	200	2210	1815	0
f_5	4	4	1954	-0.396652961085471
f_6	4	4	100336	-0.396652961085468
	8	8	130496	-0.396652961085434

The symbols are explained above. The functions are:

$f_1(x, y) = 133 + 99x - 35y$ on $[-1, 1] \times [-1, 1]$;

$f_2(x, y) = x^2 + xy + 2y^2$ on $[-1, 3] \times [-2, 4]$;

$f_3(x, y) = 100(y - x^2)^2 + (1 - x)^2$ on $[-1.2, 1.2] \times [-1.2, 1.2]$;

$f_4(x, y) = f_3(y, x)$ on the same domain;

$f_5(x, y) = \sin(x)\cos(y)\exp(-(x^2 + y^2))$ on $[-1, 2] \times [-1, 2]$;

$f_6(x, y) = f_5(x, y)$ on $[-2, 4] \times [-2, 4]$.

Comments on Table 8.1

The results for the simple functions f_1 and f_2 are hardly surprising. As expected from the behavior of procedure *glomin* on functions of one variable (see Sections 5 and 6), the number of function evaluations (N) increases with M_x and M_y.

$f_3(x, y) = 100(y - x^2)^2 + (1 - x)^2$ is the well-known Rosenbrock function (Rosenbrock (1960)), and it has a steep curved valley along the parabola $y = x^2$. Note that f_4 is just the Rosenbrock function in disguise, and it is interesting that only 1815 function evaluations were required to minimize f_4, compared to 13320 for f_3. Thus, it can make a large difference whether we minimize first over x (with y fixed) and then over y, or vice versa, but it is difficult to give a reliable rule as to which should be done first. Of course, even the lower figure of 1815 function evaluations is very high by comparison with 100 or less for methods which seek local minima (see Chapter 7), but perhaps this is the price which must be paid to guarantee that we do have the global minimum. (This is only a conjecture, for the results of Section 5 have not been extended to functions of several variables.)

The functions f_5 and f_6 are the same, but the domain of f_6 is four times as large as the domain of f_5. For this function the size of the domain has much more influence on N do the bounds M_x and M_y: increasing the size of the domain by a factor of four increased N by a factor of about 50, but doubling M_x and M_y only increased N by about 30 percent. With a different function, though, we could easily reach the opposite conclusion.

To summarize: if it is possible to give upper bounds M_x and M_y on the partial derivatives f_{xx} and f_{yy}, then procedure *glomin2d* will find a guaranteed good approximation to the global minimum, but a considerable number of function evaluations may be required if the domain of f is large or if the bounds M_x and M_y are weak. As for one-dimensional minimization, the size of the tolerance t has a fairly small influence on the total number of function evaluations required.

Finally, we should note that we have restricted ourselves to rectangular domains merely for the sake of simplicity: there is no essential difficulty in dealing with nonrectangular domains.

Section 9
SUMMARY AND CONCLUSIONS

In Section 1 we show that the problem of finding the global minimum $\varphi_f = f(\mu_f)$ of a function f defined on a compact set is well-posed, whereas the problem of finding μ_f is not well-posed. Some *a priori* conditions on f are necessary to ensure finding the global minimum, and several possible conditions are discussed in Section 1. We concentrate our attention mainly on one such condition, a given upper bound on f'', and small variations of this condition.

An efficient algorithm for one-dimensional global minimization, based on theorems in Sections 2 and 7, is described in Section 3. The effect of rounding errors, and the number of function evaluations required, are discussed in Sections 3 to 5, and numerical results are given in Section 6. Finally, in Section 8 the results for functions of one variable are used to give an algorithm for finding the global minimum of a function of several variables (practically useful for two or three variables), and ALGOL procedures are given in Section 10. The ALGOL procedures are guaranteed to give correct results, provided the basic arithmetic operations are performed with a small relative error. (See the remark following equation (3.30).)

For practical problems, the main difficulty in using the results of this chapter lies in finding the necessary bounds on second derivatives. One intriguing idea is that, if $f(\mathbf{x})$ is expressed in terms of elementary functions, then the second derivatives can be computed symbolically, and upper bounds can then be obtained from the symbolic second derivatives via simple inequalities. Thus, the entire process of finding the global minimum can be automated. In some cases functions defined on unbounded domains can also be dealt with automatically by using suitable elementary transformations.

Section 10
ALGOL 60 PROCEDURES

The ALGOL procedures *glomin* (for global minimization of a function of one variable) and *glomin2d* (for global minimization of a function of two variables) are given below. The algorithms and some numerical results are described in Sections 3 to 6 and 8. A FORTRAN translation of procedure *glomin* is given in the Appendix.

real procedure *glomin* $(a, b, c, m, macheps, e, t, f, x)$;
value $a, b, c, m, macheps, e, t$;
real $a, b, c, m, macheps\ e, t, x$; **real procedure** f;
    begin comment:
        glomin returns the global minimum value of the function $f(x)$ defined on $[a, b]$. The procedure assumes that $f \in C^2[a, b]$ and $f''(x) \leq m$ for all $x \in [a, b]$ (weaker conditions are sufficient: see Section 7). e and t are positive tolerances: we assume that $f(x)$ is computed with an absolute error bounded by e, i.e., that $|fl(f(x(1 \pm macheps))) - f(x)| \leq e$, where *macheps* is the relative machine precision. Then x and *glomin* are returned so that $\min(f) \leq f(x) \leq \min(f) + t + 2e$ and $\min(f) - e \leq glomin = fl(f(x)) \leq \min(f) + t + e$.

c is an initial guess at x (a or b will do). The number of function evaluations required is usually close to the least possible, provided t is not unreasonably small (see Sections 3 to 5);

integer k; **real** $a0$, $a2$, $a3$, $d0$, $d1$, $d2$, h, $m2$, p, q, qs, r, s, y, $y0$, $y1$, $y2$, $y3$, yb, $z0$, $z1$, $z2$;

comment: Initialization;

$x := a0 := b$; $a2 := a$;

$yb := y0 := f(b)$; $y := y2 := f(a)$;

if $y0 < y$ **then** $y := y0$ **else** $x := a$;

if $m > 0 \wedge a < b$ **then**

    begin comment: Nontrivial case ($m > 0$, $a < b$);

    $m2 := 0.5 \times (1 + 16 \times macheps) \times m$;

    if $c \le a \vee c \ge b$ **then** $c := 0.5 \times (a + b)$;

    $y1 := f(c)$; $k := 3$; $d0 := a2 - c$; $h := 9/11$;

    if $y1 < y$ **then**

        begin $x := c$; $y := y1$ **end**;

    comment: Main loop;

    next: $d1 := a2 - a0$; $d2 := c - a0$;

    $z2 := b - a2$; $z0 := y2 - y1$; $z1 := y2 - y0$;

    $p := r := d1 \times d1 \times z0 - d0 \times d0 \times z1$;

    $q := qs := 2 \times (d0 \times z1 - d1 \times z0)$;

    comment: Try to find a lower value of f using quadratic interpolation;

    if $k > 100000 \wedge y < y2$ **then go to** skip;

    retry: **if** $q \times (r \times (yb - y2) + z2 \times q \times ((y2 - y) + t))$

           $< z2 \times m2 \times r \times (z2 \times q - r)$ **then**

        begin $a3 := a2 + r/q$; $y3 := f(a3)$;

        if $y3 < y$ **then**

           begin $x := a3$; $y := y3$

           end

        end;

    comment: With probability about 0.1 do a random search for a lower value of f. Any reasonable random number generator can be used in place of the one here (it need not be very good);

    skip: $k := 1611 \times k$; $k := k - 1048576 \times (k \div 1048576)$;

    $q := 1$; $r := (b - a) \times (k/100000)$;

    if $r < z2$ **then go to** retry;

    comment: Prepare to step as far as possible;

    $r := m2 \times d0 \times d1 \times d2$; $s := \text{sqrt}(((y2 - y) + t)/m2)$;

    $h := 0.5 \times (1 + h)$;

    $p := h \times (p + 2 \times r \times s)$; $q := r + 0.5 \times qs$;

    $r := -0.5 \times (d0 + (z0 + 2.01 \times e)/(d0 \times m2))$;

    $r := a2 + (\textbf{if } r < s \vee d0 < 0 \textbf{ then } s \textbf{ else } r)$;

comment: It is safe to step to r, but we may try to step further;
$a3: = $ **if** $p \times q > 0$ **then** $a2 + p/q$ **else** r;
inner: **if** $a3 < r$ **then** $a3: = r$;
if $a3 \geq b$ **then**
  begin $a3: = b$; $y3: = yb$ **end**
else $y3: = f(a3)$;
if $y3 < y$ **then**
  begin $x: = a3$; $y: = y3$ **end**;
$d0: = a3 - a2$;
if $a3 > r$ **then**
  begin comment: Inspect the parabolic lower bound on f in
  $(a2, a3)$;
  $p: = 2 \times (y2 - y3)/(m \times d0)$;
  if $\mathrm{abs}(p) < (1 + 9 \times macheps) \times d0$
  $\wedge\ 0.5 \times m2 \times (d0 \times d0 + p \times p) >$
  $(y2 - y) + (y3 - y) + 2 \times t$ **then**
    begin comment: Halve the step and try again;
    $a3: = 0.5 \times (a2 + a3)$; $h: = 0.9 \times h$; **go to** inner
    end
  end;
 if $a3 < b$ **then**
  begin comment: Prepare for the next step;
  $a0: = c$; $c: = a2$; $a2: = a3$;
  $y0: = y1$; $y1: = y2$; $y2: = y3$;
  go to next
  end
 end;
$glomin: = y$
end glomin;

real procedure *glomin2d* $(ax, ay, bx, by, mx, my, macheps, e, t, f, x, y)$;
value $ax, ay, bx, by, mx, my, macheps, e, t$;
real $ax, ay, bx, by, mx, my, macheps, e, t, x, y$;
real procedure f;
 begin comment:
  Glomin2d returns the global minimum $z = f(x, y)$ of the function
 $f(x, y)$ defined on the rectangle $[ax, bx] \times [ay, by]$. mx and my are upper
 bounds on the second partial derivatives of f: we assume that $f_{xx}(x, y)$
 $\leq mx$ and $f_{yy}(x, y) \leq my$ in the rectangle. e and t are positive tolerances:
 f must be evaluated to an accuracy of $\pm e$, and on return
 $\min(f) \leq f(x, y) \leq \min(f) + t + 3e$ and
 $\min(f) - e \leq z = fl(f(x, y)) \leq \min(f) + t + 2e$.
 macheps is the relative machine precision, and procedure *glomin* (for

one-dimensional minimization) is assumed to be global;

real procedure *phi* (*y*); **value** *y*; **real** *y*;

      begin comment: Returns min $f(x, y)$ over x (y fixed), and may

                           alter the global variables *first*, *xs* and *zm*;

      real procedure $fx(x)$; **value** x; **real** x;

          begin fx: $= f(x, y)$ **end** fx;

      real *ym*;

      ym: $= glomin(ax, bx, xs, mx, macheps, e, t1, fx, xs)$;

      if *first* $\lor$ *ym* $<$ *zm* **then**

          begin *first*: $=$ **false**; *zm*: $=$ *ym*; *x*: $=$ *xs* **end**;

      phi: $=$ *ym*

      end *phi*;

real *t*1, *xs*, *zm*; **Boolean** *first*;

first: $=$ **true**; *zm*: $= 0$;

*t*1: $= 0.5 \times t$; *xs*: $= ax$;

glomin2d: $= glomin$ (*ay*, *by*, *ay*, *my*, *macheps*, *t*1 $+ e$, *t*1, *phi*, *y*)

end *glomin2d*;

7

A NEW ALGORITHM FOR MINIMIZING A FUNCTION OF SEVERAL VARIABLES WITHOUT CALCULATING DERIVATIVES

Section 1
INTRODUCTION AND SURVEY OF THE LITERATURE

In this chapter we consider the general unconstrained minimization problem: given a function $f: R^n \to R$, find an approximate local minimum of f. There is no need to emphasize the practical importance of this problem, and the recent literature on the subject is quite extensive. Here we give only a brief introduction, and no attempt is made to duplicate the survey articles by Box (1966), Fletcher (1965, 1969c), and Powell (1970a, e), or the books by Beale (1968); Box, Davies, and Swann (1969); Fletcher (1969a); Jacoby, Kowalik, and Pizzo (1971); Kowalik and Osborne (1968); Wilde (1964); and Wilde and Beightler (1967).

In practical problems the global minimum, not a mere local minimum, is usually of interest. Methods for finding global minima are discussed in Chapter 6, but for functions of a moderate or large number of variables the methods of Chapter 6 are impractical. Usually the best that we can do, in the absence of any special knowledge about f, is to use a good local minimizer and try several different combinations of starting positions, steplengths, etc., in the hope that the best local minimum found is the global minimum.

Constrained problems

It often happens that we want to minimize $f(\mathbf{x})$ subject to the constraint that $\mathbf{x}$ is in some subset D of R^n. (Sometimes f is only defined on D.) Simple upper and/or lower bounds, of the form

$$a_i \leq x_i \leq b_i \tag{1.1}$$

on the components x_i of $\mathbf{x}$, are particularly common, and problems with such constraints can be reduced to unconstrained problems by a transformation of variables (Box (1966)).

More general constraints may be of the form

$$g_i(\mathbf{x}) = 0 \quad \text{(an equality constraint)} \tag{1.2}$$

or

$$g_i(\mathbf{x}) \geq 0 \quad \text{(an inequality constraint)}, \tag{1.3}$$

where $g_i : D_i \subseteq R^n \to R$ is some given function, for $i = 1, \ldots, m$. $g_i(\mathbf{x})$ may be linear, say

$$g_i(\mathbf{x}) = \mathbf{a}_i^T \mathbf{x} + c_i \tag{1.4}$$

for some $\mathbf{a}_i \in R^n$ and $c_i \in R$, or $g_i(\mathbf{x})$ may be nonlinear, and perhaps quite difficult to compute. From the point of view of efficiency, it is probably best to deal with linear constraints directly, but this is difficult for nonlinear constraints. Direct methods for linear constraints are given in Fletcher (1968b), Goldfarb (1969), and Rosen (1960). (See also Bartels and Golub (1969); Bartels, Golub, and Saunders (1970); Gill and Murray (1970); Goldfarb and Lapidus (1968); Hanson (1970); and Shanno (1970b).)

Problems with nonlinear constraints can be reduced to a sequence of unconstrained problems by the use of penalty or barrier functions. See Carroll (1961); Fiacco (1961, 1969); Fiacco and Jones (1969); Fiacco and McCormick (1968); Fletcher (1969a,b); Kowalik, Osborne, and Ryan (1969); Lootsma (1968, 1970); Murray (1969); Osborne and Ryan (1970, 1971); Pietrzykowski (1969); and Zangwill (1967b). Attempts have also been made to deal with nonlinear constraints directly: see Allran and Johnsen (1970); Box (1965); Fletcher (1969a); Haarhoff and Buys (1970); Luenberger (1970); Mitchell and Kaplan (1968); Rosen (1961); and Zoutendijk (1960).

Methods using derivatives

Many methods for the constrained or unconstrained minimization of $f: D \to R$ explicitly use the partial derivatives $\partial f / \partial x_i$, for $i = 1, \ldots, n$, and some methods also use the second partial derivatives of f. Methods for constrained minimization may also use the partial derivatives of the constraint functions. An example of a derivative method is the classical method of steepest descent (Akaike (1959), Cauchy (1847), Curry (1944), Forsythe (1968), Goldstein (1962, 1965), and Ostrowski (1966, 1967a)), which repeatedly minimizes f in the direction $-\mathbf{g}$, where

$$\mathbf{g} = \begin{pmatrix} \partial f/\partial x_1 \\ \cdot \\ \cdot \\ \cdot \\ \partial f/\partial x_n \end{pmatrix} \tag{1.5}$$

is the gradient of f. Perhaps the most successful methods using derivatives are the Davidon–Fletcher–Powell "variable metric" method (Davidon (1959), Dixon (1971a,b), Fletcher and Powell (1963), Huang (1970), and McCormick (1969)), and the "conjugate gradient" method of Fletcher and Reeves (1964), which is slower but requires less storage than the variable metric method. For other methods using derivatives, and related topics, see Bard (1968, 1970); Broyden (1965, 1967, 1970a,b); Cantrell (1969); Cragg and Levy (1969); Crowder and Wolfe (1971); Daniel (1967a,b,1970); Davidon (1968, 1969); Fletcher (1966, 1970); Goldfarb (1970); Goldfeld, Quandt, and Trotter (1968); Goldstein and Price (1967); Greenstadt (1967, 1970); Luenberger (1969b); Matthews and Davies (1971); McCormick and Pearson (1969); Miele and Cantrell (1969); Myers (1968); Pearson (1969); Powell (1969b, 1970b, c, d); Ramsay (1970); Shanno and Kettler (1969); Sorensen (1969); Takahashi (1965); and Wells (1965).

In many practical problems it is difficult or impossible to find the partial derivatives of $f(\mathbf{x})$ directly. One possibility is to compute derivatives numerically, and then use one of the methods requiring derivatives. Stewart (1967) has successfully modified the variable metric method so that difference approximations to derivatives can be used. The difficulty is in balancing the influence of rounding errors and truncation errors when using finite differences to estimate derivatives. For a computer program, see Lill (1970).

Methods not using derivatives

Stewart's modification of the variable metric method appears to work well in most practical cases (see Stewart (1967), Powell (1970a), and Section 7), but it is more natural to use a method which does not need derivatives if derivatives can only be found numerically. In Chapter 5 we showed that, for one-dimensional problems, such methods can be more efficient than methods which approximate derivatives numerically, although it is not clear whether the same applies in n dimensions.

Several methods which do not use derivatives have been compared in the survey papers of Box (1966), Fletcher (1965, 1969c), Powell (1970a, e), and Spang (1962). (See also Bell and Pike (1966); Berman (1969); Box (1957); Chazan and Miranker (1970); Hooke and Jeeves (1961); Kowalik and Osborne (1968); Nelder and Mead (1965); Smith (1962); Spendley, Hext, and Himsworth (1962); Swann (1964); and Winfield (1967).) Excluding Stewart's method, the most successful method appears to be that of Powell (1964), described in Section 3. The main object of this chapter is to present some modifications which improve the speed and reliability of Powell's method. The modifications are discussed in Sections 4 to 6, and some numerical results are given in Section 7.

Quadratic convergence

Suppose that $f(\mathbf{x})$ has continuous second derivatives

$$f_{ij} = \frac{\partial^2 f}{\partial x_i \, \partial x_j} \tag{1.6}$$

for $i, j = 1, \ldots, n$, in a neighborhood N of a local minimum $\boldsymbol{\mu}$. Since $\boldsymbol{\mu}$ is a minimum, the gradient of f vanishes at $\boldsymbol{\mu}$, and the Hessian matrix

$$A = (f_{ij}) \tag{1.7}$$

is positive definite or semi-definite. Near $\boldsymbol{\mu}$, the quadratic function

$$Q(\mathbf{x}) = f(\boldsymbol{\mu}) + \tfrac{1}{2}(\mathbf{x} - \boldsymbol{\mu})^T A(\mathbf{x} - \boldsymbol{\mu}) \tag{1.8}$$

is a good approximation to $f(\mathbf{x})$. Thus, any minimization method which has ultimate fast convergence for a general function $f(\mathbf{x})$ with continuous second derivatives must have fast convergence for a positive definite quadratic function, and we might expect the converse to hold too. This observation has led to the investigation of methods which have *quadratic convergence*, i.e., which find the minimum of a positive definite quadratic function in a finite number of function and/or derivative evaluations (apart from the effect of rounding errors). Examples of methods with quadratic convergence are those of Davidon–Fletcher–Powell, Fletcher and Reeves, and Powell (1964). (This is not quite true: see Section 3.) The method of steepest descent exhibits only linear convergence on a quadratic function, so it is not quadratically convergent.

A few methods which are not quadratically convergent do exhibit superlinear convergence on quadratic forms. Examples are the methods of Rosenbrock (as modified by Davies, Swann, and Campey: see Swann (1964)); Goldstein and Price (1967); and Greenstadt (1970). There is no apparent reason why such methods should fail to perform as well as quadratically convergent methods on nonquadratic functions. Thus, quadratic convergence is a desirable property, but it is neither necessary nor sufficient for a good minimization method.

Stability: the descent property

In many methods for unconstrained minimization $f(\mathbf{x})$ has been evaluated at $\mathbf{x}_0$, the current best estimate of the position of the minimum of $f(\mathbf{x})$. A new estimate, $\mathbf{x}_1^*$, is made on the basis of the values of f at $\mathbf{x}_0$ and a small number of other points (previous best estimates, or points close to $\mathbf{x}_0$). Additional information built up from previous iterations, e.g., an approximation to the Hessian matrix of f at $\mathbf{x}_0$, may also be used. The prediction $\mathbf{x}_1^*$ may be unreliable, and it may happen that

$$f(\mathbf{x}_1^*) > f(\mathbf{x}_0). \tag{1.9}$$

For example, this often occurs if $\mathbf{x}_0$ is not close to a local minimum and an inadequate quadratic approximation to $f(\mathbf{x})$ is used.

To avoid the possibility of instability, most procedures do not accept $\mathbf{x}_1^*$ as the next approximation to the minimum. Instead, they perform a "linear search" in the direction $\mathbf{x}_1^* - \mathbf{x}_0$, i.e., they take the point

$$\mathbf{x}_1 = \mathbf{x}_0 + \lambda_0(\mathbf{x}_1^* - \mathbf{x}_0) \tag{1.10}$$

as the next approximation, where λ_0 is chosen to minimize the function

$$\varphi(\lambda) = f(\mathbf{x}_0 + \lambda(\mathbf{x}_1^* - \mathbf{x}_0)) \tag{1.11}$$

of one variable. This ensures that

$$f(\mathbf{x}_1) \leq f(\mathbf{x}_0), \tag{1.12}$$

so the successive points generated must lie in the "level set"

$$S = \{\mathbf{x} \in R^n \,|\, f(\mathbf{x}) \leq f(\mathbf{x}_0)\}. \tag{1.13}$$

In practice, it is not worthwhile to try to minimize the function $\varphi(\lambda)$ very accurately. In fact, the minimum may not even exist: $\varphi(\lambda)$ may be monotonic increasing or decreasing, or have a maximum but no minimum. Box (1966) gives examples where an attempt to minimize $\varphi(\lambda)$ too accurately prevents a minimization procedure from finding the desired minimum. It is sometimes stated that the quadratic convergence property of certain methods depends on $\varphi(\lambda)$ being minimized exactly, but all that is really required for these methods is that the one-dimensional minimization procedure minimizes a *quadratic* function of λ exactly. Thus, for quadratic convergence it is sufficient to fit a parabola $P(\lambda)$ to $\varphi(\lambda)$, and take $\lambda_0 = \lambda_0^*$, where λ_0^* minimizes $P(\lambda)$. Because of the danger of instability, this simple procedure is not acceptable, but it is reasonable to take $\lambda_0 = \lambda_0^*$ provided that

$$\varphi(\lambda_0^*) \leq \varphi(0), \tag{1.14}$$

which ensures that (1.12) holds. (Powell (1970e) gives some reasons for requiring rather more than (1.14).) See also Sections 6 and 7.

Sums of squares

A very common unconstrained minimization problem is to minimize a function $f(\mathbf{x})$ of the form

$$f(\mathbf{x}) = \sum_{i=1}^{m} [f_i(\mathbf{x})]^2, \tag{1.15}$$

for some (generally nonlinear) functions $f_i(\mathbf{x})$. For example, this problem arises when parameters $x_1, \ldots, x_n$ are fitted by the method of least squares, using m observations. An important special case occurs when the minimum value of $f(\mathbf{x})$ is zero: then we have a solution of the system of equations

$$f_i(\mathbf{x}) = 0 \tag{1.16}$$

for $i = 1, \ldots, m$.

Applying a general function minimizer to $f(\mathbf{x})$ may not be the most efficient way to minimize (1.15). Methods which make use of the individual residuals $f_i(\mathbf{x})$ are likely to be considerably more efficient than methods which merely try to minimize $f(\mathbf{x})$ without considering the individual residuals, at least if the minimum value of $f(\mathbf{x})$ is close to zero. Methods which make use of the residuals are described in Barnes (1965), Box (1966), Brent (1971a, b), Brown and Dennis (1968, 1971a, b), Brown and Conte (1967), Broyden (1965, 1969), Dennis (1968, 1969a, b), Fletcher (1968a), Gauss (1809), Hartley (1961), Jones (1970), Levenberg (1944), Marquardt (1963), Ortega and Rheinboldt (1970), Peckham (1970), Powell (1965, 1968b, 1969a), Rall (1966, 1969), Schubert (1970), Shanno (1970a), Späth (1967), Voigt (1971), Wolfe (1959), and Zeleznik (1968). Good numerical methods for solving linear least squares problems are also relevant: see Björck (1967a, b, 1968), Businger and Golub (1965), Golub (1965), Golub and Reinsch (1970), Golub and Saunders (1969), Golub and Wilkinson (1966), Hanson and Dyer (1971), and Stoer (1971).

Let us see why it may be worthwhile to use the residuals. Suppose that we have a good initial approximation to the minimum of $f(\mathbf{x})$, so the functions $f_i(\mathbf{x})$ can be closely represented by linear approximations in the region of interest. To find a linear approximation to $f_i(\mathbf{x})$, we need to evaluate $f_i(\mathbf{x})$ at $n + 1$ points, or evaluate $f_i(\mathbf{x})$ and the n components of its gradient at one point. Thus, after the same amount of work as is required for $n + 1$ evaluations of $f(\mathbf{x})$, or one evaluation of $f(\mathbf{x})$ and its gradient, the solution of a linear least squares problem gives an approximation to the minimum. This approximation is usually good if the minimum value of $f(\mathbf{x})$ is small (Powell (1965)), unless the linear problem is very ill-conditioned. On the other hand, if the special form (1.15) of $f(\mathbf{x})$ is disregarded, then it is necessary to evaluate $f(\mathbf{x})$ at $\frac{1}{2}(n + 1)(n + 2)$ points to find an approximating quadratic form. (Alternatively, f and its gradient must be evaluated at $\lceil \frac{1}{2}(n + 2) \rceil$ or more points.) This suggests that methods which disregard the special form of $f(\mathbf{x})$ are likely to be much slower than methods which use the individual residuals, at least if n is large. Empirical evidence supports this conclusion (see particularly Table 3 of Box (1966)), although some of the present methods which make use of the residuals appear to be rather unreliable.

Despite our conclusion, most of the test functions given in Section 7 are of the form (1.15). This is because a particularly simple way to construct test functions with bounded level sets is to use functions of the form (1.15).

Some additional references

The following general references on function minimization and related topics should also be mentioned: Abadie (1970); Balakrishnan (1970); Bennett (1965); Colville (1968); Dold and Eckmann (1970a, b); Evans and Gould (1970); Hadley (1964); Künzi, Tzschach, and Zehnder (1968); Lavi and Vogl (1966); Luenberger (1969a); Mangasarian (1969); Murtagh and Sargent (1970); Powell (1966, 1969c); Ralston and Wilf (1960); Rice (1970); Rosen and Suzuki (1965); Shah, Buehler, and Kempthorne (1964); Wolfe (1963, 1969, 1971); Zadeh (1969); Zangwill (1969a, b); and Zoutendijk (1966).

Section 2
THE EFFECT OF ROUNDING ERRORS

Rounding errors in the computation of $f(\mathbf{x})$ limit the accuracy attainable with any minimization method using only the computed values of $f(\mathbf{x})$. In this section we generalize the results of Section 5.2, where the same problem is considered for functions of one variable. As in Section 5.2, the results of this section do not apply to methods which use the gradient of f, computed analytically. (They do apply if the gradient is computed by finite differences.)

Suppose that, in a neighborhood N of a local minimum $\mathbf{\mu}$, the partial derivatives $f_{ij}(\mathbf{x})$ are Lipschitz continuous, i.e., for all $\mathbf{x}, \mathbf{y} \in N$,

$$| f_{ij}(\mathbf{x}) - f_{ij}(\mathbf{y}) | \leq M_{ij} \| \mathbf{x} - \mathbf{y} \|, \tag{2.1}$$

where M_{ij} is a Lipschitz constant $(i, j = 1, \ldots, n)$, and any of the usual vector norms may be used. Since the gradient of $f(\mathbf{x})$ vanishes at $\mathbf{\mu}$, a simple extension of Lemma 2.3.1 shows that, for $\mathbf{x} \in N$,

$$f(\mathbf{x}) = f(\mathbf{\mu}) + \tfrac{1}{2}(\mathbf{x} - \mathbf{\mu})^T A(\mathbf{x} - \mathbf{\mu}) + R(\mathbf{x}), \tag{2.2}$$

where

$$A = (f_{ij}(\mathbf{\mu})) \tag{2.3}$$

is the Hessian matrix of $f(\mathbf{x})$ at $\mathbf{\mu}$, and

$$| R(\mathbf{x}) | \leq M \| \mathbf{x} - \mathbf{\mu} \|^3, \tag{2.4}$$

for some constant M depending on n, the Lipschitz constants M_{ij}, and the norm.

As in Section 5.2, the best that can be expected is that the computed value $fl(f(\mathbf{x}))$ of $f(\mathbf{x})$ satisfies the nearly attainable bound

$$fl(f(\mathbf{x})) = f(\mathbf{x})(1 + \epsilon_\mathbf{x}) \qquad (2.5)$$

where

$$|\epsilon_\mathbf{x}| \leq \epsilon, \qquad (2.6)$$

and ϵ is the relative machine precision (Section 4.2). If f is computed using single-precision arithmetic, the error bound will probably be considerably worse than this.

Let δ be the largest number such that, according to equations (2.2) to (2.6), it is possible that

$$fl(f(\mathbf{\mu} + \delta\mathbf{u})) \leq f(\mathbf{\mu}), \qquad (2.7)$$

for some unit vector $\mathbf{u}$. Then it is unreasonable to expect any minimization procedure, based on single-precision evaluations of f, to return an approximation $\hat{\mathbf{\mu}}$ to $\mathbf{\mu}$ with a guaranteed upper bound for $\| \hat{\mathbf{\mu}} - \mathbf{\mu} \|$ less than δ.

Let the eigenvalues of A be $\lambda_1 \geq \lambda_2 \geq \ldots \geq \lambda_n$, with a set of corresponding normalized eigenvectors $\mathbf{u}_1, \mathbf{u}_2, \ldots, \mathbf{u}_n$. Since $\mathbf{\mu}$ is a local minimum of $f(\mathbf{x})$, certainly

$$\lambda_n \geq 0, \qquad (2.8)$$

and we suppose that $\lambda_n > 0$. (The position of the minimum is less well determined if $\lambda_n = 0$.) If $M\delta/\lambda_n$ is small compared to unity, and we take $\mathbf{u} = \mathbf{u}_n$, then (2.7) is possible with

$$\delta \simeq \sqrt{\frac{2 | f(\mathbf{\mu}) | \epsilon}{\lambda_n}}. \qquad (2.9)$$

Thus, an upper bound on $\| \hat{\mathbf{\mu}} - \mathbf{\mu} \|$ can hardly be less than the right side of (2.9).

The condition number

With the assumptions above, and δ given by (2.9),

$$f(\mathbf{\mu} + \delta\mathbf{u}_1) \simeq f(\mathbf{\mu}) + \kappa\epsilon \, | f(\mathbf{\mu}) |, \qquad (2.10)$$

where

$$\kappa = \frac{\lambda_1}{\lambda_n} \qquad (2.11)$$

is the spectral condition number of A. We shall say that κ is the condition number of the minimization problem for the local minimum $\mathbf{\mu}$. The condition number determines the rate of convergence of some minimization methods (e.g., steepest descent), and it is also important because rounding errors make it difficult to solve problems with condition numbers of the order of ϵ^{-1} or greater.

Scaling

A change of scale along the coordinate axes has the effect of replacing the Hessian matrix A by SAS, where S is a positive diagonal matrix. The problem of choosing S to minimize the condition number of SAS is difficult, even if A is known explicitly. (See Forsythe and Moler (1967) for the problem of minimizing the condition number of $S_1 A S_2$, where A is not necessarily symmetric.) A good general rule is that SAS should be roughly row (and hence column) equilibrated: see Wilkinson (1963, 1965a). In practical minimization problems little is known about the Hessian matrix until a reasonable approximation to the minimum has been found. This suggests that a scale-dependent function minimizer could incorporate an automatic scaling procedure, using current information about A to determine the scaling. One way to do this is described in Section 4.

Section 3
POWELL'S ALGORITHM

In this section we briefly describe Powell's algorithm for minimization without calculating derivatives. The algorithm is described more fully in Powell (1964), and a small error in this paper is pointed out by Zangwill (1967a). Numerical results are given in Fletcher (1965), Box (1966), and Kowalik and Osborne (1968). A modified algorithm, which is suitable for use on a parallel computer, and which converges for strictly convex C^2 functions with bounded level sets, is described by Chazan and Miranker (1970).

Powell's method is a modification of a quadratically convergent method proposed by Smith (1962). Both methods ensure convergence in a finite number of steps, for a positive definite quadratic function, by making use of some properties of conjugate directions.

Conjugate directions

If A is positive definite and symmetric, then minimizing the quadratic function

$$\mathbf{x}^T A \mathbf{x} - 2\mathbf{b}^T \mathbf{x} = (\mathbf{x} - A^{-1}\mathbf{b})^T A (\mathbf{x} - A^{-1}\mathbf{b}) - \mathbf{b}^T A^{-1}\mathbf{b} \qquad (3.1)$$

is equivalent to solving the system of linear equations

$$A\mathbf{x} = \mathbf{b}. \qquad (3.2)$$

If the matrix A is known explicitly then, instead of minimizing (3.1), we can solve (3.2) by any suitable method: for example, by forming the Cholesky decomposition of A. In the applications of interest here, A is the Hessian matrix of a certain function, and is not known explicitly, but the equivalence of the problems (3.1) and (3.2) is still useful.

DEFINITION 3.1

Two vectors $\mathbf{u}$ and $\mathbf{v}$ are said to be *conjugate* with respect to the positive definite symmetric matrix A if

$$\mathbf{u}^T A \mathbf{v} = 0. \tag{3.3}$$

When there is no risk of confusion, we shall simply say that $\mathbf{u}$ and $\mathbf{v}$ are conjugate. By a set of *conjugate directions*, we mean a set of vectors which are pairwise conjugate.

Remark

If $\{\mathbf{u}_1, \ldots, \mathbf{u}_m\}$ is any set of *nonzero* conjugate directions in R^n, then $\mathbf{u}_1, \ldots, \mathbf{u}_m$ are linearly independent. Thus $m \leq n$; and $m = n$ iff $\mathbf{u}_1, \ldots, \mathbf{u}_m$ span R^n.

THEOREM 3.1

If A is positive definite symmetric, $A\mathbf{x} = \mathbf{b}$, and $\{\mathbf{u}_1, \ldots, \mathbf{u}_m\}$ is a set of nonzero conjugate directions, then

$$\mathbf{y} = \mathbf{x} - \sum_{i=1}^{m} \left(\frac{\mathbf{u}_i^T \mathbf{b}}{\mathbf{u}_i^T A \mathbf{u}_i} \right) \mathbf{u}_i \tag{3.4}$$

is conjugate to each of $\mathbf{u}_1, \ldots, \mathbf{u}_m$.

Proof

If $1 \leq j \leq m$ then, from (3.4),

$$\mathbf{u}_j^T A \mathbf{y} = \mathbf{u}_j^T (A\mathbf{x} - \mathbf{b}) = 0. \tag{3.5}$$

COROLLARY 3.1

If $m = n$ in Theorem 3.1, then $\mathbf{y} = \mathbf{0}$, so

$$\mathbf{x} = \sum_{i=1}^{n} \left(\frac{\mathbf{u}_i^T \mathbf{b}}{\mathbf{u}_i^T A \mathbf{u}_i} \right) \mathbf{u}_i. \tag{3.6}$$

Returning to the minimization problem, Theorem 3.1 and the equivalence of problems (3.1) and (3.2) give the following result.

THEOREM 3.2

If A is positive definite symmetric,

$$f(\mathbf{x}) = \mathbf{x}^T A \mathbf{x} - 2\mathbf{b}^T \mathbf{x} + c \tag{3.7}$$

for some $\mathbf{b} \in R^n$ and $c \in R$, and $\mathbf{u}_1, \ldots, \mathbf{u}_m$ is a set of nonzero conjugate directions, then the minimum of $f(\mathbf{x})$ in the space spanned by $\mathbf{u}_1, \ldots, \mathbf{u}_m$

occurs at the point $\sum_{i=1}^{m} \beta_i \mathbf{u}_i$, where

$$\beta_i = \frac{\mathbf{u}_i^T \mathbf{b}}{\mathbf{u}_i^T A \mathbf{u}_i}. \tag{3.8}$$

Proof

This follows from Theorem 3.1 or, alternatively, from the relation

$$f\left(\sum_{i=1}^{m} \alpha_i \mathbf{u}_i\right) = \sum_{i=1}^{m} (\alpha_i - \beta_i)^2 \mathbf{u}_i^T A \mathbf{u}_i + c - \sum_{i=1}^{m} \frac{(\mathbf{u}_i^T \mathbf{b})^2}{\mathbf{u}_i^T A \mathbf{u}_i} \tag{3.9}$$

(cross terms vanish because of the conjugacy of $\mathbf{u}_1, \ldots, \mathbf{u}_m$).

The usefulness of Theorem 3.2 stems from the following result, which shows how we can calculate the β_i of (3.8) using function evaluations, even if A, $\mathbf{b}$, and c are not known explicitly. The proof is immediate from equation (3.9).

THEOREM 3.3

With the notation of Theorem 3.2, a fixed j satisfying $1 \leq j \leq m$, and fixed $\alpha_1, \ldots, \alpha_{j-1}, \alpha_{j+1}, \ldots, \alpha_m$, the minimum of

$$\varphi_j(\alpha_j) = f\left(\sum_{i=1}^{m} \alpha_i \mathbf{u}_i\right) \tag{3.10}$$

occurs at $\alpha_j = \beta_j$.

From Theorems 3.2 and 3.3, we see that the minimum of the quadratic function $f(\mathbf{x})$ can be found by n one-dimensional minimizations along nonzero conjugate directions $\mathbf{u}_1, \ldots, \mathbf{u}_n$, and the order in which the one-dimensional minimizations are performed is irrelevant. To use this result, we have to be able to generate sets of conjugate directions. Both Powell's method and Smith's method do this by using the following theorem, given in Powell (1964).

THEOREM 3.4

If the minimum of $f(\mathbf{x})$ (given by (3.7)) in the direction $\mathbf{u}$ from the point $\mathbf{x}_i^*$ is at $\mathbf{x}_i$, for $i = 0, 1$, then $\mathbf{x}_1 - \mathbf{x}_0$ is conjugate to $\mathbf{u}$.

Proof

For $i = 0$ and 1,

$$\frac{\partial}{\partial \lambda} f(\mathbf{x}_i + \lambda \mathbf{u}) = 0 \tag{3.11}$$

at $\lambda = 0$, so, from (3.7),

$$\mathbf{u}^T(A\mathbf{x}_i - \mathbf{b}) = 0. \tag{3.12}$$

Subtracting equations (3.12) for $i = 0$ and 1 gives

$$\mathbf{u}^T A(\mathbf{x}_1 - \mathbf{x}_0) = 0, \tag{3.13}$$

which completes the proof.

Powell's basic procedure

We can now describe the basic idea of Powell's algorithm. Let x_0 be the initial approximation to the minimum, and let $u_1, \ldots, u_n$ be the columns of the identity matrix. One iteration of the basic procedure consists of the following steps:

1. For $i = 1, \ldots, n$, compute β_i to minimize $f(x_{i-1} + \beta_i u_i)$, and define $x_i = x_{i-1} + \beta_i u_i$.
2. For $i = 1, \ldots, n - 1$, replace u_i by u_{i+1}.
3. Replace u_n by $x_n - x_0$.
4. Compute β to minimize $f(x_0 + \beta u_n)$, and replace x_0 by $x_0 + \beta u_n$.

For a general (non-quadratic) function, the iteration is repeated until some stopping criterion is satisfied. If f is quadratic, consider the situation after the k-th iteration, where $1 \leq k \leq n$. Then $u_{n-k+1}, \ldots, u_n$ are conjugate, by Theorem 3.4 and the choice of u_n at step 3: see Powell (1964). After n iterations we have minimized along n conjugate directions $u_1, \ldots, u_n$ and, by Theorems 3.2 and 3.3, the minimum has been reached if the u_i are all nonzero. This is true if $\beta_1 \neq 0$ at each iteration, for then the directions $u_1, \ldots, u_n$ cannot become linearly dependent.

The problem of linear dependence

Zangwill (1967a) observed that, even for a quadratic function f, one of the iterations may have $\beta_1 = 0$. This results in the directions $u_1, \ldots, u_n$ becoming linearly dependent, and from then on the procedure can only find the minimum of $f(x)$ over a proper subspace of R^n. Even though it is unlikely that β_1 will vanish exactly, Powell discovered that the directions $u_1, \ldots, u_n$ often become nearly linearly dependent. Thus, he suggested that the new direction $x_n - x_0$ should be used, and one of the old $u_1, \ldots, u_n$ discarded, only if this does not decrease the value of $|\det(v_1, \ldots, v_n)|$, where

$$v_i = (u_i^T A u_i)^{-1/2} u_i \tag{3.14}$$

for $i = 1, \ldots, n$. With this modification the algorithm is quite successful (see Fletcher (1965) and Box (1966) for a comparison with other methods), but the desirable property of quadratic convergence is lost, for a complete set of conjugate directions may never be built up. In the next section, we describe a different way of avoiding the problem of linear dependence of the search directions. The numerical results given in Section 7 suggest that our method of ensuring linear independence may be preferable to Powell's. Zangwill (1967a) suggested a simpler way of ensuring independence, but numerical experiments (Rhead (1971)) show that Powell's modification is preferable to Zangwill's.

Section 4

THE MAIN MODIFICATION

The simplest way to avoid linear dependence of the search directions with Powell's basic procedure, and retain quadratic convergence if $\beta_1 \neq 0$, is to reset the search directions $\mathbf{u}_1, \ldots, \mathbf{u}_n$ to the columns of the identity matrix after every n or $n + 1$ iterations. A similar "restarting" device is suggested by Fletcher and Reeves (1964) for their conjugate gradient method, and restarting is actually necessary to ensure superlinear convergence (Crowder and Wolfe (1971)). For other methods, restarting may slow down convergence, because information built up about the function is periodically thrown away.

Instead of resetting $U = [\mathbf{u}_1, \ldots, \mathbf{u}_n]$ to the identity matrix, we can equally well reset U to any orthogonal matrix Q. To avoid discarding useful information about f, we choose Q so that $\mathbf{u}_1, \ldots, \mathbf{u}_n$ remain conjugate if f is quadratic. Principal vectors $\mathbf{q}_1, \ldots, \mathbf{q}_n$ are computed on the assumption that f is quadratic, and U is reset to $Q = [\mathbf{q}_1, \ldots, \mathbf{q}_n]$. The motivation for this procedure may be summarized thus:

1. If the quadratic approximation to f is good, then the new search directions are conjugate with respect to a matrix which is close to the Hessian matrix of f at the minimum, and thus subsequent iterations give fast convergence.

2. Regardless of the validity of the quadratic approximation, the new search directions are orthogonal, so the search for a minimum can never become restricted to a subspace.

The extra computation involved

We show below that finding principal axes does not require any extra function evaluations, but it does involve finding an orthogonal set of eigenvectors for a symmetric matrix H of order n. This requires about $6n^3$ multiplications, and a similar number of additions, if done as suggested below. Since the principal axes are found only once for every n^2 linear minimizations, and a linear minimization requires about 2.25 function evaluations on the average (see Section 7), the extra computation is less than $3n$ multiplications per function evaluation. We can expect the evaluation of a nontrivial function of n variables to require considerably more than $3n$ multiplications, so the overhead caused by our modification is not excessive. Also, it may be worth paying a little for the principal axis reduction, for the extra information about f is often of interest. For example, it shows the sensitivity of $f(\mathbf{x})$ to slight changes in $\mathbf{x}$ near the minimum. The principal axes and eigenvalues may be useful in statistical problems: see Nelder and Mead (1965).

Finding the principal vectors

Suppose that

$$f(\mathbf{x}) = \mathbf{x}^T A \mathbf{x} - 2\mathbf{b}^T \mathbf{x} + c \qquad (4.1)$$

is a positive definite quadratic function, although A, $\mathbf{b}$, and c may not be known explicitly. If n iterations of Powell's basic procedure are performed as described above, and at each iteration $\beta_1 \neq 0$, then we obtain n nonzero conjugate directions $\mathbf{u}_1, \ldots, \mathbf{u}_n$. Let $U = [\mathbf{u}_1 \ldots \mathbf{u}_n]$. By the conjugacy of $\mathbf{u}_1, \ldots, \mathbf{u}_n$,

$$U^T A U = D, \qquad (4.2)$$

where D is a diagonal matrix with positive diagonal elements d_i.

During the last (i.e., n-th) iteration, we have performed one-dimensional minimizations in the directions $\mathbf{u}_1, \ldots, \mathbf{u}_n$. Consider a minimization from the point $\mathbf{x}_{i-1}$, in the direction $\mathbf{u}_i$, for $1 \leq i \leq n$. We minimize the function

$$\varphi_i(\alpha) = f(\mathbf{x}_{i-1} + \alpha \mathbf{u}_i) \qquad (4.3)$$

$$= \alpha^2 \mathbf{u}_i^T A \mathbf{u}_i + 2\alpha(\mathbf{u}_i^T A \mathbf{x}_{i-1} - \mathbf{u}_i^T \mathbf{b}) + (\mathbf{x}_{i-1}^T A \mathbf{x}_{i-1} - 2\mathbf{x}_{i-1}^T \mathbf{b} + c). \qquad (4.4)$$

To minimize $\varphi_i(\alpha)$ we fit a parabola, which necessitates computing the second difference $\varphi_i[\alpha_0, \alpha_1, \alpha_2]$ for three distinct points α_0, α_1, and α_2. From equation (4.4),

$$\varphi_i[\alpha_0, \alpha_1, \alpha_2] = \mathbf{u}_i^T A \mathbf{u}_i = d_i, \qquad (4.5)$$

so the diagonal elements d_i of D are known without any extra computation. (If the quadratic approximation to $\varphi_i(\alpha)$ is bad and $\varphi_i[\alpha_0, \alpha_1, \alpha_2] \leq 0$, then we arbitrarily set d_i to a small positive number.)

Let

$$V = U D^{-1/2} \qquad (4.6)$$

be the matrix with columns $\mathbf{v}_1, \ldots, \mathbf{v}_n$ given by (3.14), and let

$$H = A^{-1}. \qquad (4.7)$$

Since U is nonsingular, equation (4.2) gives

$$H = U D^{-1} U^T = V V^T. \qquad (4.8)$$

The matrix V is easily computed from U in n^2 multiplications and n square roots, but the computation of $V V^T$ is more expensive, and can be avoided: see below.

Our aim is to find the principal axes of the quadratic function f, i.e., to find an orthogonal matrix Q such that

$$Q^T A Q = \Lambda, \qquad (4.9)$$

where $\Lambda = \text{diag}(\lambda_i)$ is diagonal. Thus, the columns $\mathbf{q}_i$ of Q are just the eigenvectors of A, with corresponding eigenvalues $\lambda_1, \ldots, \lambda_n$, and we can assume that $\lambda_1 \geq \cdots \geq \lambda_n$. The obvious way to find Q and Λ is to compute $H = VV^T$ explicitly, and then find Q and Λ such that

$$Q^T H Q = \Lambda^{-1} \qquad (4.10)$$

by finding the eigensystem of H.

Use of the singular value decomposition to find Q and Λ

If the condition number $\kappa = \lambda_1/\lambda_n$ is of order ϵ^{-1}, where ϵ is the relative machine precision (Section 4.2), then rounding errors may lead to disastrous errors in the computed small eigenvalues $\lambda_1^{-1}, \lambda_2^{-1}, \ldots$ of H, and in the corresponding eigenvectors $\mathbf{q}_1, \mathbf{q}_2, \ldots$, even if they are well-determined by V. Thus, it may be necessary to compute H, and find its eigensystem, using double-precision arithmetic. This difficulty can be avoided if, instead of forming $H = VV^T$, we work directly with V. Suppose that we find the singular value decomposition of V, i.e., find orthogonal matrices Q and R such that

$$Q^T V R = \Sigma, \qquad (4.11)$$

where $\Sigma = \mathrm{diag}(\sigma_i)$ is a diagonal matrix. (See Golub and Kahan (1965), and Kogbetliantz (1955).) Then

$$\Lambda^{-1} = Q^T H Q = (Q^T V R)(Q^T V R)^T = \Sigma^2, \qquad (4.12)$$

so Q is the desired matrix of eigenvectors of A, and the eigenvalues λ_i are given by

$$\lambda_i = \sigma_i^{-2}. \qquad (4.13)$$

Note that the matrix R is not required, and it is not necessary to compute VV^T.

Since it is desirable that the computed matrix Q should be close to an orthogonal matrix, we suggest that Q and Σ should be found by the method of Golub and Reinsch (1970). This involves reducing V to bidiagonal form by Householder transformations (Parlett (1971)), and then computing the singular value decomposition of the bidiagonal matrix by a variant of the QR algorithm.

Let us compare the amount of computational work involved in computing Q and Λ via

1. The singular value decomposition (SVD) of V as described above, and

2. Finding the matrix H and its eigensystem, using Householder's reduction to tridiagonal form and then the QR algorithm. (See Bowdler, Martin, Reinsch, and Wilkinson (1968); Francis (1962); Householder (1964); Kublanovskaya (1961); Martin, Reinsch, and Wilkinson (1968); and Wilkinson (1965a, b, 1968).)

For purposes of comparison we count only multiplications, and ignore terms of order n^2. We also suppose that the QR process requires pn iterations, for some modest number p.

For method 1, the Householder reduction requires $4n^3/3$ multiplications, accumulation of the left-hand transformations requires another $4n^3/3$ multiplications, and the QR process with accumulation of the transformations requires $2pn^3$ multiplications if no splitting occurs. Thus, method 1 requires $(8 + 6p)n^3/3$ multiplications in all.

For method 2, the Householder reduction requires $2n^3/3$ multiplications (only half as much as for method 1 because of symmetry), accumulation of the transformations requires $2n^3/3$ multiplications, and the QR process requires $2pn^3$, giving $(4 + 6p)n^3/3$ altogether. This could be reduced to $4n^3/3$, still ignoring terms of order n^2, if inverse iteration were used to compute the eigenvectors of the tridiagonal matrix, but then it would be difficult to guarantee orthogonality of eigenvectors corresponding to close or multiple eigenvalues. Another $\frac{1}{2}n^3$ multiplications are needed to compute $H = VV^T$ by the usual method (but taking advantage of symmetry), making $(11 + 12p)n^3/6$ multiplications in all.

The ratio of the work involved for methods 1 and 2 is thus

$$r = \frac{16 + 12p}{11 + 12p} < \frac{16}{11}, \tag{4.14}$$

and for a typical value of $p = 1.6$ we have $r \simeq 1.17$. Thus, method 1 can be expected to be only about 20 percent slower than the numerically inferior method 2. Both methods require temporary storage for only a few n-vectors, apart from the n by n matrix V which is overwritten by Q.

Automatic scaling

We mentioned in Section 2 that a general minimization procedure could incorporate automatic scaling of the independent variables, in an attempt to reduce the condition number of the problem. Scaling has the effect of replacing the matrix V above by $S^{-1}V$, where S is a positive diagonal matrix. The ALGOL procedure *praxis* of Section 9 chooses S automatically to try to reduce the condition number of $S^{-1}V$. S is chosen so that $S^{-1}V$ is row-equilibrated, with the constraint that

$$1 \leq s_{ii} \leq scbd, \tag{4.15}$$

where *scbd* is a bound which may be set to 1 if no scaling is desired. If $scbd = \infty$, then our algorithm (like Powell's) is independent of scale changes, except for the stopping criterion. Numerical experiments on the examples described in Section 7 suggest that *scbd* should be fairly small (about 10) unless the axes are very badly scaled initially. The automatic scaling is worthwhile, but it may be unreliable, which is the reason for introducing *scbd*. Thus, the user should not forget to try to scale his problem as well as possible.

Another modification

For Powell's basic procedure to minimize a positive definite quadratic function in n iterations, steps 1 to 3 of the first iteration are unnecessary. Thus, our algorithm omits steps 1 to 3 on the first iteration, and also after each singular value decomposition (i.e., at the $(n + 1)$-st, $(2n + 1)$-st, ... iterations). Thus there are exactly $1 + (n - 1)(n + 1) = n^2$ linear minimizations, instead of $n(n + 1)$, between successive singular value decompositions. This modification is not important for large n, but numerical results suggest that it is worthwhile for small n.

Section 5
THE RESOLUTION RIDGE PROBLEM

Suppose temporarily that we are trying to *maximize* a function $f(\mathbf{x}_1, \mathbf{x}_2)$ of two variables by an ascent method. Wilde (1964) points out that rounding errors in the computation of f may lead to premature termination because of the "resolution ridge" problem illustrated in Diagram 5.1.

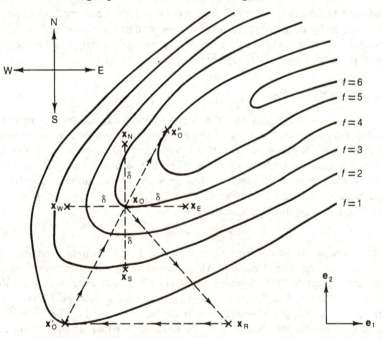

DIAGRAM 5.1   **A resolution ridge**

Regard the surface defined by $f(x_1, x_2)$ as a hill. We may reach a point $\mathbf{x}_0$, situated on a narrow ridge, and then try to proceed to a higher point by performing linear searches in certain directions. Suppose, for example, that we attempt linear searches in the *EW* and *NS* directions. The point $\mathbf{x}_0$ is not at the true maximum of f in both these directions but, because of the effect of rounding errors in evaluating f, our one-dimensional search procedure will only attempt to locate the position of maxima to within some positive tolerance δ (see Section 2). Let $\mathbf{x}_E = \mathbf{x}_0 + \delta\mathbf{e}_1$, $\mathbf{x}_W = \mathbf{x}_0 - \delta\mathbf{e}_1$, $\mathbf{x}_N = \mathbf{x}_0 + \delta\mathbf{e}_2$, and $\mathbf{x}_S = \mathbf{x}_0 - \delta\mathbf{e}_2$. It may happen that $f(\mathbf{x}_0)$ is greater than each of $f(\mathbf{x}_N)$, $f(\mathbf{x}_S)$, $f(\mathbf{x}_E)$, and $f(\mathbf{x}_W)$, so $\mathbf{x}_0$ is within the tolerance δ of local maxima in both of the search directions, even though $\mathbf{x}_0$ is a long way from the true maximum, which could be reached by climbing up the ridge. The same problem can arise with functions of more than two variables, or if we are looking for a minimum rather than a maximum (when we might speak of a "resolution valley" problem).

It is clear from the diagram that, if we know another point $\mathbf{x}_0'$ on the ridge, then a linear search in the direction $\mathbf{x}_0 - \mathbf{x}_0'$ will give a point $\mathbf{x}_0''$ with $f(\mathbf{x}_0'') > f(\mathbf{x}_0)$ unless the ridge is sharply curved. This is one motivation for the method suggested by Rosenbrock (1960), and improved by Davies, Swann, and Campey. (See Swann (1964), and also Andrews (1969), Baer (1962), Fletcher (1965, 1969c, d), Osborne (1969), Palmer (1969), Powell (1968a), and Section 7.)

Finding another point on the ridge

If linear searches from the point $\mathbf{x}_0$ fail to give a higher point, and a resolution ridge is suspected, then the following strategy may be successful: take a step of length about 10δ in a random direction from $\mathbf{x}_0$, reaching the point $\mathbf{x}_R$. Then perform one or more linear searches, starting at $\mathbf{x}_R$, and reaching the point $\mathbf{x}_0'$. As Diagram 5.1 shows, the point $\mathbf{x}_0'$ is likely to be on the ridge, so a linear search in the direction $\mathbf{x}_0 - \mathbf{x}_0'$ may now be successful.

Although he does not refer to the resolution ridge problem, Powell (1964) uses such a strategy as part of his stopping criterion. We propose to use this strategy during the regular iterations as well.

Incorporating a random step into Powell's basic procedure

Suppose that we are commencing iteration k of Powell's basic procedure, and $2 \leq k \leq n$. To ensure quadratic convergence, we must search along the directions $\mathbf{u}_{n-k+2}, \dots, \mathbf{u}_n$ in step 1 of iteration k, but the searches along directions $\mathbf{u}_1, \dots, \mathbf{u}_{n-k+1}$ are not necessary for quadratic convergence. (They are desirable for other reasons: see Fletcher (1965) for a comparison of Powell's method and Smith's method.) The quadratic convergence property still holds if, at step 1, we move to any point

$$\mathbf{x}_{n-k+1} = \mathbf{x}_0 + \sum_{i=1}^{n} \beta_i' \mathbf{u}_i \tag{5.1}$$

with $\beta_1' \neq 0$, before performing linear searches in the directions $\mathbf{u}_{n-k+2}$, ..., $\mathbf{u}_n$. Thus, before performing linear searches in directions $\mathbf{u}_1, \ldots, \mathbf{u}_n$ at step 1 of iteration k, we may try the random step strategy as described above. Procedure *praxis* does this if the problem appears to be ill-conditioned or if recent linear searches have failed to improve the current approximation to the minimum.

This modification is not necessary for well-conditioned problems, but numerical results show that it is essential in order to ensure that a good approximation to the minimum is found for very ill-conditioned problems. For example, consider minimizing

$$f(\mathbf{x}) = \mathbf{x}^T A \mathbf{x}, \tag{5.2}$$

where A is a ten by ten Hilbert matrix (i.e., $a_{ij} = 1/(i+j-1)$ for $1 \leq i, j \leq 10$), with a condition number of 1.6×10^{13}. Using an IBM 360 computer with machine precision 16^{-13}, and starting from $(1, 1, \ldots, 1)^T$, our algorithm successfully found the position of the minimum of $f(\mathbf{x})$ to within the specified tolerance of 10^{-5}, but it failed without the random step strategy. (For further details, see Section 7.)

Extrapolation along the valley

If the function minimizer has been descending a valley for several complete cycles, the quadratic approximation to f is obviously inadequate (or the minimum would already have been found), and it may be worthwhile to try an extrapolation along the valley. Suppose that, immediately before three successive singular value decompositions, the best approximations to the minimum are $\mathbf{x}'$, $\mathbf{x}''$, and $\mathbf{x}'''$, with $d_0 = \|\mathbf{x}' - \mathbf{x}''\|_2 > 0$ and $d_1 = \|\mathbf{x}'' - \mathbf{x}'''\|_2 > 0$. Numerical tests indicate that curved valleys are often approximated fairly well by the space-curve

$$\mathbf{x}(\lambda) = \frac{\lambda(\lambda - d_1)}{d_0(d_0 + d_1)}\mathbf{x}' - \frac{(\lambda + d_0)(\lambda - d_1)}{d_0 d_1}\mathbf{x}'' + \frac{\lambda(\lambda + d_0)}{d_1(d_0 + d_1)}\mathbf{x}''',$$

$$\tag{5.3}$$

which satisfies $\mathbf{x}(-d_0) = \mathbf{x}'$, $\mathbf{x}(0) = \mathbf{x}''$, and $\mathbf{x}(d_1) = \mathbf{x}'''$. Before the third, fourth, fifth, ... singular value decompositions, procedure *praxis* (Section 9) moves to the point $\mathbf{x}(\lambda_0)$, where λ_0 approximately minimizes $f(\mathbf{x}(\lambda))$. λ_0 is computed by the procedure that performs linear searches.

Section 6
SOME FURTHER DETAILS

In this section we give some more details of the ALGOL procedure *praxis* of Section 9. The criterion for discarding search directions, the linear search procedure, and the stopping criterion are described briefly.

The discarding criterion

Suppose for the moment that $f(\mathbf{x})$ is the quadratic function given by equation (3.7). In steps 2 and 3 of Powell's basic procedure (Section 3), we effectively discard the search direction $\mathbf{u}_1$ and replace it by $\mathbf{x}_n - \mathbf{x}_0$. The algorithm suggested by Powell does not necessarily discard $\mathbf{u}_1$: instead, as mentioned in Section 3, it discards one of $\mathbf{u}_1, \ldots, \mathbf{u}_n, \mathbf{u}_{n+1} = \mathbf{x}_n - \mathbf{x}_0$, so as to maximize

$$|\det(\mathbf{v}_1 \ldots \mathbf{v}_n)|, \tag{6.1}$$

where $\mathbf{v}_i$ is given by equation (3.14) after renumbering the remaining n directions. We wish to retain convergence for a quadratic form in n iterations, so we are not free to discard any one of $\mathbf{u}_1, \ldots, \mathbf{u}_{n+1}$. At the k-th iteration, for $2 \le k \le n$, we can discard any one of $\mathbf{u}_1, \ldots, \mathbf{u}_{n-k+1}$ without losing quadratic convergence (see Section 5). For lack of a better criterion, we choose to discard the direction, from $\mathbf{u}_1, \ldots, \mathbf{u}_{n-k+1}$, to maximize the resulting determinant (6.1).

Suppose that the new direction $\mathbf{x}_n - \mathbf{x}_0 = \mathbf{u}_{n+1}$ satisfies

$$\frac{\mathbf{u}_{n+1}}{(\mathbf{u}_{n+1}^T A \mathbf{u}_{n+1})^{1/2}} = \sum_{i=1}^{n} \alpha_i \frac{\mathbf{u}_i}{(\mathbf{u}_i^T A \mathbf{u}_i)^{1/2}}. \tag{6.2}$$

The effect of discarding $\mathbf{u}_i$, replacing it by $\mathbf{u}_{n+1}$, and then renumbering the directions, is to multiply the determinant (6.1) by $|\alpha_i|$. Thus, our criterion is to choose i, with $1 \le i \le n - k + 1$, so that $|\alpha_i|$ is maximized. If $\beta_1, \ldots, \beta_n$ are as in the description of Powell's basic procedure (Section 3), and the linear minimization with step $\beta_i \mathbf{u}_i$ decreases $f(\mathbf{x})$ by an amount Δ_i, then, from (3.7),

$$\Delta_i = \beta_i^2 \mathbf{u}_i^T A \mathbf{u}_i, \tag{6.3}$$

so $\sqrt{\Delta_i}/|\beta_i|$ may be used as an estimate of $(\mathbf{u}_i^T A \mathbf{u}_i)^{1/2}$. (If $\beta_i = 0$ we use the result of a previous iteration.)

Suppose that the random step procedure described in Section 5 moves from $\mathbf{x}_0$ to

$$\mathbf{y}_0 = \mathbf{x}_0 + \sum_{i=1}^{n} \gamma_i \mathbf{u}_i \tag{6.4}$$

before the linear searches in the directions $\mathbf{u}_1, \ldots, \mathbf{u}_n$ are performed. Then

$$\mathbf{u}_{n+1} = \mathbf{x}_n - \mathbf{x}_0 = \sum_{i=1}^{n} (\beta_i + \gamma_i)\mathbf{u}_i, \tag{6.5}$$

and the β'_i of equation (5.1) are given by

$$\beta'_i = \begin{cases} \beta_i + \gamma_i & \text{if } 1 \le i \le n - k + 1, \\ \gamma_i & \text{if } n - k + 2 \le i \le n. \end{cases} \tag{6.6}$$

From (6.2), (6.3), and (6.5),

$$(\mathbf{u}_{n+1}^T A \mathbf{u}_{n+1})^{1/2}\alpha_i = \frac{(\beta_i + \gamma_i)\sqrt{\Delta_i}}{|\beta_i|}, \tag{6.7}$$

so we must discard direction $\mathbf{u}_i$, with $1 \le i \le n - k + 1$, to maximize the modulus of the right side of (6.7). Since this does not explicitly depend on the matrix A, the same criterion is used even if f is not necessarily quadratic. Note that our criterion reduces to Powell's, apart from our restriction that $i \le n - k + 1$, if there are no random steps (i.e., if $\gamma_i = 0$ for $i = 1, \ldots, n$). Quadratic convergence is guaranteed if we ensure that, for $k = 2, \ldots, n$,

$$\beta'_1 = \beta'_2 = \cdots = \beta'_{n-k+1} = 0 \tag{6.8}$$

never holds at iteration k.

The linear search

Our linear search procedure is similar to that suggested by Powell (1964). We wish to find a value of λ which approximately minimizes

$$\varphi(\lambda) = f(\mathbf{x}_0 + \lambda\mathbf{u}), \tag{6.9}$$

where the initial point $\mathbf{x}_0$ and direction $\mathbf{u} \ne \mathbf{0}$ are given, and $\varphi(0) = f(\mathbf{x}_0)$ is already known. If a linear search in the direction $\mathbf{u}$ has already been performed, or if $\mathbf{u}$ resulted from a singular value decomposition, then an estimate of $\varphi''(0)$ is available. A parabola $P(\lambda)$ is fitted to $\varphi(\lambda)$, using $\varphi(0)$, the estimate of $\varphi''(0)$ if available, and the computed value of $\varphi(\lambda)$ at another point, or at two points if there is no estimate of $\varphi''(0)$. If $P(\lambda)$ has a minimum at $\lambda = \lambda^*$, and $\varphi(\lambda^*) < \varphi(0)$, then λ^* is accepted as a value of λ to minimize (6.9) approximately. Otherwise λ^* is replaced by $\lambda^*/2$, $\varphi(\lambda^*)$ is re-evaluated, and the test is repeated. (After a number of unsuccessful tries, the procedure returns with $\lambda = 0$.)

The stopping criterion

The user of procedure *praxis* provides two parameters: t (a positive absolute tolerance), and $\epsilon = macheps$ (the machine precision). The procedure attempts to return $\mathbf{x}$ satisfying

$$\|\mathbf{x} - \boldsymbol{\mu}\|_2 \le \epsilon^{1/2}\|\mathbf{x}\|_2 + t, \tag{6.10}$$

where μ is the position of the true local minimum near $\mathbf{x}$. The exact form of the right side of (6.10) is not important, and could easily be changed. It was chosen because of the analogy with the one-dimensional case (Chapter 5).

It is impossible to guarantee that (6.10) will hold for all functions f, or even for f which are C^2 near μ. Our stopping criterion is, however, rather cautious, and (6.10) is satisfied for all but one of the numerical examples discussed in Section 7. The sole exception is the extremely ill-conditioned problem

$$f(\mathbf{x}) = \mathbf{x}^T A \mathbf{x}, \tag{6.11}$$

where A is a twelve by twelve Hilbert matrix with condition number $\kappa \simeq 1.7 \times 10^{16} > \epsilon^{-1} \simeq 4 \times 10^{15}$. In most cases the stopping criterion is over-cautious, and some unnecessary function evaluations are performed. We remark, as does Powell (1964), that the stopping criterion is not an essential part of our algorithm. An improved criterion could easily be incorporated.

Let $\mathbf{x}'$ be the current best approximation to the minimum before an iteration of the basic procedure, and let $\mathbf{x}''$ be the best approximation after the iteration. We test if

$$2 \| \mathbf{x}' - \mathbf{x}'' \|_2 \leq \epsilon^{1/2} \| \mathbf{x}'' \|_2 + t. \tag{6.12}$$

The stopping criterion is simply to stop, and return the approximation $\mathbf{x}''$, if (6.12) is satisfied for a prescribed number of consecutive iterations. The number of consecutive iterations depends on how cautious we wish to be: two is reasonable, and was used for the examples described in Section 7. Because the random step strategy described in Section 5 is always used if (6.12) was satisfied on the previous iteration, there is no need for a more complicated criterion such as the one used by Powell (1964).

Section 7
NUMERICAL RESULTS AND COMPARISON WITH OTHER METHODS

The ALGOL W procedure *praxis*, given in Section 9, has been tested on IBM 360/67 and 360/91 computers with machine precision 16^{-13}. In this section we summarize the results of the numerical tests, and compare them with results for other methods reported in the literature. Our procedure has also been translated into SAIL (an extension of ALGOL: see Swinehart and Sproull (1970)), and used to solve least-squares parameter-fitting problems with up to 16 variables on a PDP 10 computer with machine precision 2^{-26}. The parameter-fitting problem is described in Sobel (1970).

Table 7.1 summarizes the performance of procedure *praxis* on the test functions described below. In all cases the tolerance $t = 10^{-5}$ and *macheps* $= 16^{-13}$. The table gives the number of variables, n; the initial step-size (a rough estimate of the distance to the minimum), h; and the starting point,

$\mathbf{x}_0$. So that the results can be compared with those of methods with a different stopping criterion, we give the number n_f of function evaluations and the number n_l of linear searches (including any parabolic extrapolations) required to reduce $f(\mathbf{x}) - f(\boldsymbol{\mu})$ below 10^{-10}, where $f(\boldsymbol{\mu})$ is the true minimum of f. As $f(\mathbf{x})$ was only printed out after each iteration of the basic procedure, (i.e., after every n linear minimizations), the number of function evaluations required to reduce $f(\mathbf{x}) - f(\boldsymbol{\mu})$ to 10^{-10} is usually slightly less than n_f, so we also give the actual value of $f(\mathbf{x}) - f(\boldsymbol{\mu})$ after n_f function evaluations. Finally, the table gives κ, the estimated condition number of the problem. Except for the few cases where it is easily found analytically, κ is estimated from the computed singular values, and may be rather inaccurate.

TABLE 7.1   **Results for various test functions**

Function	n	h	$\mathbf{x}_0^T$	n_f	n_l	$f(\mathbf{x}) - f(\boldsymbol{\mu})$	κ
Rosenbrock	2	1	$(-1.2, 1)$	120	47	6.61′–18	2508
Rosenbrock	2	3	$(3, 3)$	110	42	8.53′–17	2508
Rosenbrock	2	12	$(8, 8)$	181	67	9.71′–18	2508
Cube	2	1	$(-1.2, -1)$	177	68	7.18′–18	10018
Beale	2	1	$(0.1, 0.1)$	54	22	2.00′–15	162
Helix	3	1	$(-1, 0, 0)$	155	67	1.75′–11	500
Powell	3	1	$(0, 1, 2)$	55	23	1.99′–11	28
Box*	3	20	$(0, 10, 20)$	100	37	2.37′–13	8300
Singular*	4	1	$(3, -1, 0, 1)$	234	106	9.76′–11	∞
Wood*	4	10	$-(3, 1, 3, 1)$	452	191	6.06′–14	1400
Chebyquad	2	0.1	$x_i = i/(n+1)$	31	12	7.89′–20	1.3
Chebyquad	4	0.1	$x_i = i/(n+1)$	74	32	7.89′–11	7
Chebyquad	6	0.1	$x_i = i/(n+1)$	223	101	7.00′–13	50
Chebyquad	8	0.1	$x_i = i/(n+1)$	326	147	6.32′–11	200?
Watson*	6	1	0^T	316	145	2.83′–12	86000
Watson*	9	1	0^T	1184	541	3.18′–11	1.7′9
Tridiag	4	8	0^T	27	11	0	29.3
Tridiag	6	12	0^T	51	22	0	64.9
Tridiag	8	16	0^T	126	55	0	113
Tridiag	10	20	0^T	201	89	1.56′–15	175
Tridiag	12	24	0^T	259	118	2.23′–15	250
Tridiag	16	32	0^T	488	222	1.26′–13	438
Tridiag	20	40	0^T	805	379	0	677
Hilbert	2	10	$(1, \ldots, 1)$	11	4	3.98′–15	19
Hilbert	4	10	$(1, \ldots, 1)$	50	22	6.11′–15	1.5′4
Hilbert	6	10	$(1, \ldots, 1)$	133	58	1.50′–11	1.5′7
Hilbert	8	10	$(1, \ldots, 1)$	262	119	8.14′–11	1.5′10
Hilbert†	10	10	$(1, \ldots, 1)$	592	267	7.84′–11	1.6′13
Hilbert†	12	10	$(1, \ldots, 1)$	731	328	1.98′–11	1.7′16

*For these results we set *illc*: = **true** in the initialization phase of procedure *praxis*, and the random number generator was initialized by calling *raninit*(2) in procedure *test*.

†For these results the stopping criterion was more conservative: we set *ktm*: = 4 in the initialization phase of procedure *praxis*.

For those examples marked with an asterisk, the random step strategy was used from the start. (In the initialization phase of procedure *praxis*, the variable *illc* was set to **true**.) For the other examples the procedure was used as given in Section 9 (with *illc* set to **false** initially). Although the automatic scaling feature (Section 4) reduces n_f by about 25 percent for some of the badly scaled problems, this feature was switched off for the examples given in the table. (The bound *scbd* of equation (4.15) was set to 1.)

Definitions of the test functions, and comments on the results summarized in Table 7.1, are given below.

A cautionary note

When comparing different minimization methods such as ours, Powell's, and Stewart's, the reader should not forget that the numerical results reported for the methods may have been obtained on different computers (with different word-lengths), and with different linear search procedures. Except for ill-conditioned problems, the effect of different word-lengths should only be significant in the final stages of the search, when rounding errors determine the limiting accuracy attainable. This is another reason why we prefer to consider the number of function evaluations required to reduce $f(\mathbf{x}) - f(\mathbf{\mu})$ to a reasonable threshold, rather than the number required for convergence.

Because apparently minor differences in the linear search procedure can be quite important, Fletcher (1965) prefers to consider the number of linear searches, n_l, instead of the number of function evaluations, n_f. This approach discriminates against methods such as Powell's, which use most of the search directions several times, and can thus use second derivative estimates to reduce the number of function evaluations required for the second and later searches in each direction. Note that, for the examples given in Table 7.1, n_f/n_l lies between 2.1 and 2.7, but it would be at least 3.0 for methods which do not use second derivative information, if the linear search involves fitting a parabola and evaluating f at the minimum of the parabola. Also, there are promising methods which do not use linear searches at all (see Broyden (1967), Davidon (1968, 1969), Goldstein and Price (1967), and Powell (1970e)), and these methods can be adapted to accept difference approximations to derivatives. Thus, we prefer to compare methods on the basis of the number of function evaluations required, and regard the linear search procedure, if any, as an integral part of each method.

Definitions of the test functions and comments on Table 7.1

Rosenbrock (Rosenbrock (1960)):
$$f(\mathbf{x}) = 100(x_2 - x_1^2)^2 + (1 - x_1)^2. \tag{7.1}$$
This is a well-known function with a parabolic valley. Descent methods tend to fall into the valley and then follow it around to the minimum at $(1, 1)^T$.

Details of the progress of the algorithm, for the starting point $(-1.2, 1)^T$, are given in Table 7.2. In Diagram 7.1 we compare these results with those reported for Stewart's method (Stewart (1967)), Powell's method, and the method of Davies, Swann, and Campey (as reported by Fletcher (1965)). The graph shows that our method compares favorably with the other methods. Although the function (7.1) is rather artificial, similar curved valleys often arise when penalty function methods are used to reduce constrained problems to unconstrained problems: consider minimizing $(1 - x_1)^2$, with the constraint that $x_2 = x_1^2$, by a simple-minded penalty function method.

Cube (Leon (1966)):

$$f(\mathbf{x}) = 100(x_2 - x_1^3)^2 + (1 - x_1)^2. \tag{7.2}$$

This function is similar to Rosenbrock's, and much the same remarks apply. Here the valley follows the curve $x_2 = x_1^3$.

Beale (Beale (1958)):

$$f(\mathbf{x}) = \sum_{i=1}^{3} [c_i - x_1(1 - x_2^i)]^2, \tag{7.3}$$

where $c_1 = 1.5$, $c_2 = 2.25$, $c_3 = 2.625$. This function has a valley approaching the line $x_2 = 1$, and has a minimum of 0 at $(3, \frac{1}{2})^T$. Kowalik and Osborne (1968) report that the Davidon-Fletcher-Powell algorithm reduced f to 2.18×10^{-11} in 20 function and gradient evaluations (equivalent to 60 function evaluations if the usual $(n + 1)$ weighting factor is used), and Powell's method required 86 function evaluations to reduce f to 2.94×10^{-8}. Thus, our method compares favorably on this example.

Helix (Fletcher and Powell (1963)):

$$f(\mathbf{x}) = 100[(x_3 - 10\theta)^2 + (r - 1)^2] + x_3^2, \tag{7.4}$$

where

$$r = (x_1^2 + x_2^2)^{1/2} \tag{7.5}$$

and

$$2\pi\theta = \begin{cases} \arctan(x_2/x_1) & \text{if } x_1 > 0, \\ \pi + \arctan(x_2/x_1) & \text{if } x_1 < 0. \end{cases} \tag{7.6}$$

This function of three variables has a helical valley, and a minimum at $(1, 0, 0)^T$. The results are given in more detail in Table 7.3 and Diagram 7.2. For this example our method is faster than Powell's, but slightly slower than Stewart's.

Powell (Powell (1964)):

$$f(\mathbf{x}) = 3 - \left(\frac{1}{1 + (x_1 - x_2)^2}\right) - \sin\left(\frac{\pi}{2} x_2 x_3\right) - \exp\left\{\left[-\left(\frac{x_1 + x_2}{x_2}\right) - 2\right]^2\right\}. \tag{7.7}$$

For a description of this function, see Powell (1964). Perhaps by good luck, our procedure had no difficulty with this function: it found the true minimum quickly and did not stop prematurely.

Box (Box (1966)):

$$f(\mathbf{x}) = \sum_{i=1}^{10} \left\{ \begin{array}{c} [\exp(-ix_1/10) - \exp(-ix_2/10)] \\ -x_3[\exp(-i/10) - \exp(-i)] \end{array} \right\}^2. \tag{7.8}$$

This function has minima of 0 at $(1, 10, 1)^T$, and also along the line $\{(\lambda, \lambda, 0)^T\}$. (Our procedure found the first minimum.) Kowalik and Osborne (1968) report that Powell's method took 205 function evaluations to reduce f to 3.09×10^{-9}, so our method is about twice as fast. Our method took 79 function evaluations to reduce f to 2.29×10^{-7}, so it is faster, in this example, than any of the methods compared by Box (1966), with the exception of Powell's method for sums of squares (Powell (1965)). See the comment in Section 1 about special methods for minimizing sums of squares!

Singular (Powell (1962)):

$$f(\mathbf{x}) = (x_1 + 10x_2)^2 + 5(x_3 - x_4)^2 + (x_2 - 2x_3)^4 + 10(x_1 - x_4)^4. \tag{7.9}$$

This function is difficult to minimize, and provides a severe test of the stopping criterion, because the Hessian matrix at the minimum ($\mathbf{x} = \mathbf{0}$) is doubly singular. The function varies very slowly near $\mathbf{0}$ in the two-dimensional subspace $\{(10\lambda_1, -\lambda_1, \lambda_2, \lambda_2)^T\}$. Table 7.4 and Diagram 7.3 suggest that the algorithm converges only linearly, as does Powell's algorithm. It is interesting to note that the output from our procedure would strongly suggest the singularity, if we did not know it in advance: after 219 function evaluations, with $f(\mathbf{x}) = 7.67 \times 10^{-9}$, the computed eigenvalues were 101.0, 9.999, 0.003790, and 0.001014. (The exact eigenvalues at $\mathbf{0}$ are 101, 10, 0, and 0.) After 384 function evaluations, with $f(\mathbf{x})$ reduced to 1.02×10^{-17}, the two smallest eigenvalues were 1.56×10^{-7} and 5.98×10^{-8}. Thus, our procedure should allow singularity of the Hessian matrix to be detected, in the unlikely event that it occurs in a practical problem. (For one example, see Freudenstein and Roth (1963).)

Wood (Colville (1968)):

$$f(\mathbf{x}) = 100(x_2 - x_1^2)^2 + (1 - x_1)^2 + 90(x_4 - x_3^2)^2 + (1 - x_3)^2$$
$$+ 10.1[(x_2 - 1)^2 + (x_4 - 1)^2] + 19.8(x_2 - 1)(x_4 - 1). \tag{7.10}$$

This function is rather like Rosenbrock's, but with four variables instead of two. Procedures with an inadequate stopping criterion may terminate prematurely on this function (McCormick and Pearson (1969)), but our procedure successfully found the minimum at $\boldsymbol{\mu} = (1, 1, 1, 1)^T$.

Chebyquad (Fletcher (1965)):

$f(\mathbf{x})$ is defined by the ALGOL procedure given by Fletcher (1965). As the minimization problem is still valid, we have not corrected a small error in this procedure, which does not compute exactly what Fletcher intended. In contrast to most of our other test functions, which are designed to be difficult to minimize, this function is fairly easy to minimize. For $n = 1(1)7$ and 9 the minimum is 0; for other n it is nonzero. (For $n = 8$ it is approximately 0.00351687372568.) The results given below, and illustrated in Diagrams 7.4 to 7.7, show that our method is faster than those of Powell or Davies, Swann, and Campey, but a little slower than Stewart's.

Watson (Kowalik and Osborne (1968)):

$$f(\mathbf{x}) = x_1^2 + (x_2 - x_1^2 - 1)^2$$
$$+ \sum_{i=2}^{30} \left\{ \sum_{j=2}^{n} (j-1)x_j \left(\frac{i-1}{29}\right)^{j-2} - \left[\sum_{j=1}^{n} x_j \left(\frac{i-1}{29}\right)^{j-1}\right]^2 - 1 \right\}^2.$$
(7.11)

Here a polynomial

$$p(t) = x_1 + x_2 t + \cdots + x_n t^{n-1} \tag{7.12}$$

is fitted, by least squares, to approximate a solution of the differential equation

$$\frac{dz}{dt} = 1 + z^2, \ z(0) = 0, \tag{7.13}$$

for $t \in [0, 1]$. (The exact solution is $z = \tan t$.) The minimization problem is ill-conditioned, and rather difficult to solve, because of a bad choice of basis functions $\{1, t, \ldots, t^{n-1}\}$. For $n = 6$, the minimum is $f(\boldsymbol{\mu}) \simeq 2.28767005355 \times 10^{-3}$, at $\boldsymbol{\mu} \simeq (-0.015725, 1.012435, -0.232992, 1.260430, -1.513729, 0.992996)^T$. For $n = 9$, $f(\boldsymbol{\mu}) \simeq 1.399760138 \times 10^{-6}$, and $\boldsymbol{\mu} \simeq (-0.000015, 0.999790, 0.014764, 0.146342, 1.000821, -2.617731, 4.104403, -3.143612, 1.052627)^T$. (We do not claim that all the figures given are significant.)

Kowalik and Osborne (1968) report that, after 700 function evaluations, Powell's method had only reduced f to 2.434×10^{-3} (for $n = 6$), so our method is at least twice as fast here. The Watson problem for $n = 9$ is very ill-conditioned, and seems to be a good test for a minimization procedure.

Tridiag (Gregory and Karney (1969), pp. 41 and 74):

$$f(\mathbf{x}) = \mathbf{x}^T A \mathbf{x} - 2x_1, \tag{7.14}$$

where

$$A = \begin{bmatrix} 1 & -1 & & & & \\ -1 & 2 & -1 & & \text{\Large 0} & \\ & -1 & 2 & -1 & & \\ & & -1 & 2 & -1 & \\ \text{\Large 0} & & & \ldots\ldots\ldots & \\ & & & & -1 & 2 \end{bmatrix}. \tag{7.15}$$

This function is useful for testing the quadratic convergence property. The minimum $f(\mu) = -n$ occurs when μ is the first column of A^{-1}, i.e.,

$$\mu = (n, n - 1, n - 2, \ldots, 2, 1)^T. \tag{7.16}$$

The results given in Table 7.1 show that, as expected, the minimum is found in n^2 or less linear minimizations. The eigenvalues of A are just $\lambda_j = 4\cos^2[j\pi/(2n + 1)]$ for $j = 1, \ldots, n$.

Hilbert

$$f(x) = x^T A x, \tag{7.17}$$

where A is an n by n Hilbert matrix, i.e.,

$$a_{ij} = \frac{1}{i + j - 1} \tag{7.18}$$

for $1 \leq i, j \leq n$. Like (7.14), (7.17) is a positive definite quadratic function, but the condition number increases rapidly with n. Because of the effect of rounding errors, more than n^2 linear minimizations were required to reduce f to 10^{-10} for $n \geq 4$. The procedure successfully found the minimum $\mu = 0$, to within the prescribed tolerance, for $n \leq 10$. For $n = 12$, some components of the computed minimum were greater than 0.1, even though f was reduced to 2.76×10^{-15}. This illustrates how ill-conditioned the problem is!

Some more detailed results

Tables 7.2 to 7.8 give more details of the progress of our procedure (B) on the Rosenbrock, Helix, Singular, and Chebyquad functions. In Diagrams 7.1 to 7.7, we plot

$$\Delta = \log_{10}(f(x) - f(\mu)) \tag{7.19}$$

against n_f, the number of function evaluations. Using the results given by Fletcher (1965) and Stewart (1967), the corresponding graphs for the methods of Davies, Swann, and Campey (D), Powell (P), and Stewart (S) are also given, for purposes of comparison. Results for Stewart's method on Chebyquad ($n = 8$) are not available.

TABLE 7.2 Rosenbrock

n_f	n_l	$f(\mathbf{x})$	x_1	x_2
1	0	2.42′1	−1.200000	1.000000
11	4	4.14	−1.034611	1.071270
21	8	3.42	−0.811598	0.621199
31	12	2.59	−0.549031	−0.258076
45	17	1.67	−0.268211	0.046503
58	22	1.07	−0.028125	−0.010783
72	27	3.71′−1	0.482692	0.200894
84	32	2.79′−3	0.947231	0.897130
98	37	5.89′−4	0.996384	0.990382
109	42	6.69′−9	0.999991	0.999974
120	47	6.61′−18	1.000000	1.000000
132	52	1.13′−23	1.000000	1.000000
155	57	4.47′−24	1.000000	1.000000

TABLE 7.3 Helix

n_f	n_l	$f(\mathbf{x})$	x_1	x_2	x_3
1	0	2.50′3	−1.000000	0.000000	0.000000
14	5	1.62′2	1.000000	2.000000	2.000000
23	9	1.18′2	0.563832	1.952025	1.759493
36	14	5.22	0.311857	1.000020	2.096124
44	18	4.04	0.305534	0.967190	1.987145
57	23	3.78	0.347506	0.907981	1.922708
65	27	3.01	0.847973	0.734103	1.074593
82	33	9.46′−1	0.816717	0.566910	0.969820
91	37	3.66′−1	0.965734	0.342023	0.548844
105	43	2.46′−1	1.004624	0.239418	0.364506
113	47	2.84′−2	0.993843	0.091699	0.153178
126	53	6.35′−3	1.002319	0.045726	0.072132
134	57	8.01′−4	1.002726	0.002303	0.002966
147	63	8.66′−6	0.999996	0.001853	0.002942
155	67	1.75′−11	1.000000	8.49′−9	2.47′−7
169	73	1.12′−20	1.000000	−6.45′−11	−9.92′−11
178	77	1.99′−24	1.000000	−1.69′−13	−2.47′−13
200	83	1.94′−24	1.000000	−1.60′−13	−2.53′−13

TABLE 7.4   Singular*

n_f	n_l	$f(\mathbf{x})$	n_f	n_l	$f(\mathbf{x})$
1	0	2.15′2	219	99	7.67′–9
19	6	1.18′1	234	106	9.76′–11
31	11	7.96	244	111	2.03′–12
42	16	7.75	254	116	4.11′–13
58	22	2.94	269	123	2.61′–14
68	27	9.86′–1	279	128	6.43′–15
78	32	1.34′–1	289	133	8.88′–16
94	38	6.92′–3	308	140	7.35′–16
104	43	1.18′–3	319	145	3.87′–16
114	48	5.25′–5	330	150	9.92′–17
129	55	8.25′–6	358	157	9.92′–17
139	60	2.13′–6	373	162	1.65′–17
149	65	2.70′–7	384	167	1.02′–17
164	72	7.91′–8	404	174	9.95′–18
174	77	3.95′–8	421	179	6.02′–23
184	82	3.90′–8	436	184	5.89′–23
199	89	3.90′–8	464	191	5.89′–23
209	94	3.89′–8	486	196	5.89′–23

*$\hat{\boldsymbol{\mu}}^T \simeq (-9.73 \times 10^{-7}, 9.73 \times 10^{-8}, 5.31 \times 10^{-7}, 5.31 \times 10^{-7})$, lying approximately in the subspace $\{(10\lambda_1, -\lambda_1, \lambda_2, \lambda_2)\}$, as expected. See also the first comment under Table 7.1.

TABLE 7.5   Chebyquad: $n = 2$*

n_f	n_l	$f(\mathbf{x})$
1	0	1.98′–1
12	4	4.53′–3
22	8	1.89′–8
31	12	7.89′–20
45	17	4.89′–24
73	22	4.89′–24

*$\hat{\boldsymbol{\mu}}^T = (0.2113249, 0.7886751)$

TABLE 7.6   Chebyquad: $n = 4$*

n_f	n_l	$f(\mathbf{x})$
1	0	7.12′–2
17	6	1.43′–2
27	11	1.59′–3
38	16	1.00′–4
54	22	4.22′–7
64	27	1.86′–8
74	32	7.89′–11
87	38	7.75′–14
98	43	1.88′–16

*$\hat{\mu}T = (0.1026728, \quad 0.4062037, \quad 0.5937963, \\ 0.8973272)$

TABLE 7.7   Chebyquad: $n = 6$*

n_f	n_l	$f(\mathbf{x})$
1	0	4.64′–2
23	8	2.35′–2
37	15	1.80′–2
51	22	1.21′–2
66	29	5.69′–3
81	36	2.07′–3
103	44	9.89′–5
117	51	3.47′–5
131	58	2.14′–5
145	65	1.14′–5
159	72	2.71′–6
181	80	1.13′–7
195	87	6.59′–10
209	94	1.38′–10
223	101	7.00′–13
238	108	3.77′–15

*$\hat{\mu}T = (0.066877, 0.288741, 0.366682, 0.633318, \\ 0.711259, 0.933123)$

TABLE 7.8 Chebyquad: $n = 8$*

n_f	n_l	$f(\mathbf{x})$
1	0	0.0386176982859
29	10	0.0171124413073
47	19	0.0109131815974
65	28	0.0102860269896
83	37	0.0093337335931
102	46	0.0071908595069
125	55	0.0049952481593
144	64	0.0044432513463
172	74	0.0037940416125
190	83	0.0035390722159
208	92	0.0035269968747
226	101	0.0035191392494
244	110	0.0035180637576
262	119	0.0035176364629
280	128	0.0035171964541
308	138	0.0035168743745
326	147	0.0035168737890
345	156	0.0035168737290
364	165	0.0035168737288

*$\hat{\mu}^T = (0.043153, 0.193091, 0.266329, 0.500000, 0.500000, 0.733671, 0.806910, 0.956847)$

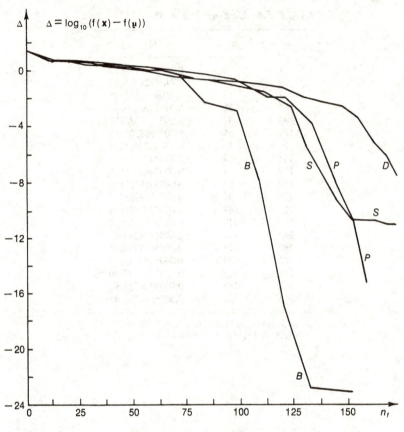

DIAGRAM 7.1 **Rosenbrock**

KEY:

    B: Our method;

    D: The method of Davies, Swann, and Campey, as given by Fletcher (1965);

    P: Powell's (1964) method, as given by Fletcher (1965);

    S: Stewart's method, as given by Stewart (1967).

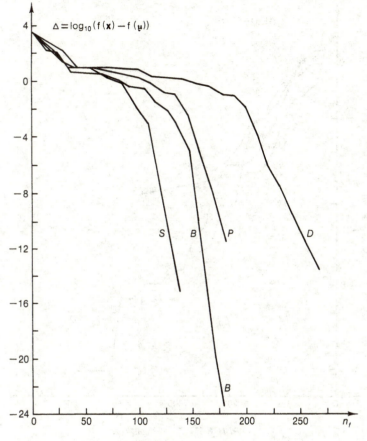

DIAGRAM 7.2 Helix

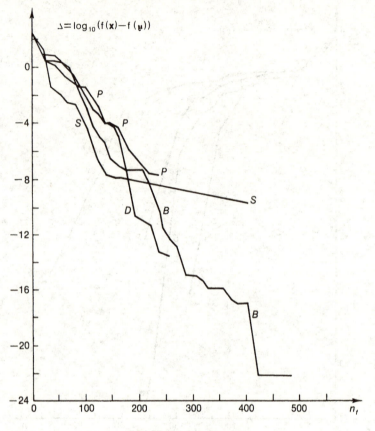

DIAGRAM 7.3   **Singular**

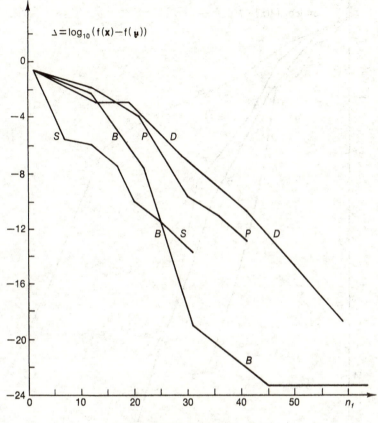

DIAGRAM 7.4   **Chebyquad,** $n = 2$

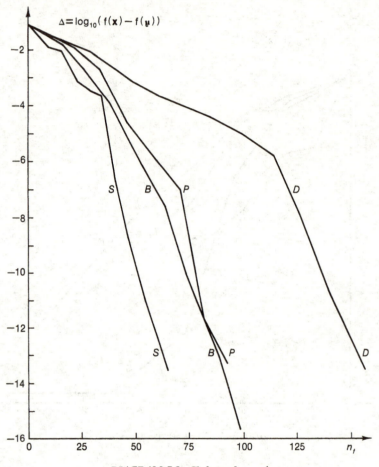

DIAGRAM 7.5 Chebyquad, $n = 4$

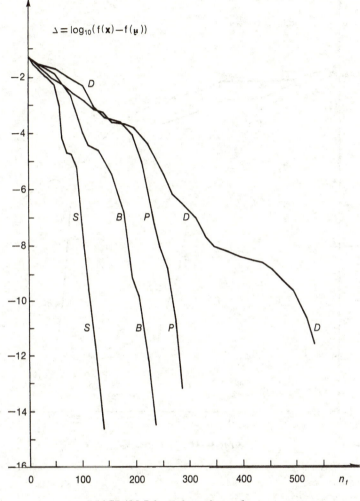

DIAGRAM 7.6   **Chebyquad,** $n = 6$

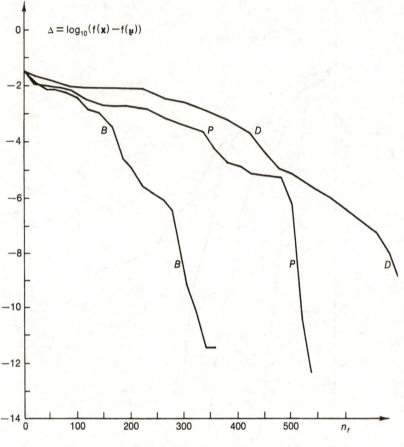

$$\Delta = \log_{10}(f(\mathbf{x}) - f(\mathbf{\mu}))$$

DIAGRAM 7.7 Chebyquad, $n = 8$

Section 8
CONCLUSION

Powell (1964) observes that, with his suggested criterion for accepting new search directions (Section 3), there is a tendency for the new directions to be accepted less often as the number of variables increases, and the quadratic convergence property of his basic procedure is lost. Our aim was to avoid this difficulty, keep the quadratic convergence property, and ensure that the search directions continue to span the whole space, while using basically the same method as Powell to generate conjugate directions.

The numerical results given in Section 7 suggest that our algorithm is faster than Powell's, and comparable to Stewart's, if the criterion is the number of function evaluations required to reduce $f(\mathbf{x})$ to a certain threshold. Also, our algorithm seems to be reliable even for very ill-conditioned problems like Watson ($n = 9$) and Hilbert ($n = 10$), while Stewart's method breaks down because of numerical difficulties on some functions, e.g., the Rosenbrock and Singular functions (see Stewart (1967)). However, we should not try to conclude too much from the numerical results: see the warning in Section 7.

Theoretical convergence results

Suppose that all arithmetic is exact, and consider our algorithm with the stopping criterion removed. Since the algorithm keeps on performing linear searches along n orthogonal directions, the same conditions that ensure convergence of the method of coordinate search will ensure convergence of our algorithm to a local minimum. In particular, the algorithm will converge to the (unique) minimum for all functions f which are C^1, strictly convex, and satisfy

$$\lim_{\lambda \to \infty} f(\lambda \mathbf{e}) = +\infty \qquad (8.1)$$

for all nonzero vectors $\mathbf{e}$. Of course, this result has limited practical interest, for in practice rounding errors may be very important: see Section 5.

It is plausible that our algorithm converges superlinearly if the Hessian matrix of f is strictly positive definite at the minimum. McCormick (1969) shows that this is true for the reset Davidon–Fletcher–Powell algorithm, provided a Lipschitz condition is satisfied. Figures 7.1, 7.2, and 7.4 to 7.7 certainly suggest that convergence is superlinear until rounding errors become important, but we do not have a proof of this conjecture: perhaps additional conditions on f, or a slight modification of the algorithm, are necessary. Some algorithms for which it is fairly easy to prove good theoretical convergence results are described in Brent (1971c).

Section 9
AN ALGOL W PROCEDURE AND TEST PROGRAM

The procedure *praxis*, with a driver program and test functions, is given below. The language is ALGOL W (Wirth and Hoare (1966); Bauer, Becker, and Graham (1968)), but none of the special features of ALGOL W have been used, so translation into another dialect of ALGOL should be straightforward.

```
BEGIN COMMENT:
              TEST PROGRAM FOR PROCEDURE PRAXIS.
              *****************************************;

LONG REAL PROCEDURE PRAXIS (LONG REAL VALUE T, MACHEPS, H;
    INTEGER VALUE N, PRIN;
    LONG REAL ARRAY X(*);  LONG REAL PROCEDURE F, RANDOM);
    BEGIN COMMENT:

    THIS PROCEDURE MINIMIZES THE FUNCTION F(X, N) OF N
    VARIABLES X(1), ... X(N), USING THE PRINCIPAL AXIS METHOD.
    ON ENTRY X HOLDS A GUESS, ON RETURN IT HOLDS THE ESTIMATED
    POINT OF MINIMUM, WITH (HOPEFULLY) |ERROR| <
    SQRT(MACHEPS)*|X| + T, WHERE MACHEPS IS THE MACHINE
    PRECISION, THE SMALLEST NUMBER SUCH THAT 1 + MACHEPS > 1,
    T IS A TOLERANCE, AND |.| IS THE 2-NORM.  H IS THE MAXIMUM
    STEP SIZE:  SET TO ABOUT THE MAXIMUM EXPECTED DISTANCE FROM
    THE  GUESS TO THE MINIMUM (IF H IS SET TOO SMALL OR TOO
    LARGE THEN THE INITIAL RATE OF CONVERGENCE WILL BE SLOW).
    THE USER SHOULD OBSERVE THE COMMENT ON HEURISTIC NUMBERS
    AFTER PROCEDURE QUAD.
    PRIN CONTROLS THE PRINTING OF INTERMEDIATE RESULTS.
    IF PRIN = 0, NO RESULTS ARE PRINTED.
    IF PRIN = 1, F IS PRINTED AFTER EVERY N+1 OR N+2 LINEAR
    MINIMIZATIONS, AND FINAL X IS PRINTED, BUT INTERMEDIATE
    X ONLY IF N <= 4.
    IF PRIN = 2, EIGENVALUES OF A AND SCALE FACTORS ARE ALSO
    PRINTED.
    IF PRIN = 3, F AND X ARE PRINTED AFTER EVERY FEW LINEAR
    MINIMIZATIONS.
    IF PRIN = 4, EIGENVECTORS ARE ALSO PRINTED.
    FMIN IS A GLOBAL VARIABLE:  SEE PROCEDURE PRINT.
    RANDOM IS A PARAMETERLESS LONG REAL PROCEDURE WHICH RETURNS
    A RANDOM NUMBER UNIFORMLY DISTRIBUTED IN (0, 1).    ANY
    INITIALIZATION MUST BE DONE BEFORE THE CALL TO PRAXIS.
    THE PROCEDURE IS MACHINE-INDEPENDENT, APART FROM THE OUTPUT
    STATEMENTS AND THE SPECIFICATION OF MACHEPS.   WE ASSUME THAT
    MACHEPS**(-4) DOES NOT OVERFLOW (IF IT DOES THEN MACHEPS MUST
    BE INCREASED), AND THAT ON FLOATING-POINT UNDERFLOW THE
    RESULT IS SET TO ZERO;

    PROCEDURE MINFIT (INTEGER VALUE N;  LONG REAL VALUE EPS, TOL;
        LONG REAL ARRAY AB(*,*);    LONG REAL ARRAY Q(*));
        BEGIN COMMENT:  AN IMPROVED VERSION OF MINFIT, SEE GOLUB &
                        REINSCH (1969), RESTRICTED TO M = N, P = 0.
                        THE SINGULAR VALUES OF THE ARRAY AB ARE
                        RETURNED IN Q, AND AB IS OVERWRITTEN WITH
                        THE ORTHOGONAL MATRIX V SUCH THAT
                        U.DIAG(Q) = AB.V,
                        WHERE U IS ANOTHER ORTHOGONAL MATRIX;
    INTEGER L, KT;
    LONG REAL C,F,G,H,S,X,Y,Z;
    LONG REAL ARRAY E(1::N);
    COMMENT: HOUSEHOLDER'S REDUCTION TO BIDIAGONAL FORM;
    G := X := 0;
    FOR I :=  1 UNTIL N DO
      BEGIN
      E(I) := G;  S := 0;  L := I+1;
      FOR J := I UNTIL N DO S := S+AB(J,I)**2;
      IF S<TOL THEN G := 0 ELSE
        BEGIN
        F := AB(I,I);  G := IF F<0 THEN LONGSQRT(S)
                                  ELSE -LONGSQRT(S);
        H := F*G-S;  AB(I,I) := F-G;
        FOR J := L UNTIL N DO
          BEGIN F := 0;
```

```
          FOR K := I UNTIL N DO F := F + AB(K,I)*AB(K,J);
          F := F/H;
          FOR K := I UNTIL N DO AB(K,J) := AB(K,J) + F*AB(K,I)
          END J
        END S;
      Q(I) := G;  S := 0;
      IF I<=N THEN FOR J := L UNTIL N DO
          S := S + AB(I,J)**2;
      IF S<TOL THEN G := 0 ELSE
        BEGIN
        F := AB(I,I+1);  G := IF F<0 THEN LONGSQRT(S)
                           ELSE -LONGSQRT(S);
        H := F*G-S;  AB(I,I+1) := F-G;
        FOR J := L UNTIL N DO E(J) := AB(I,J)/H;
        FOR J := L UNTIL N DO
          BEGIN S := 0;
          FOR K := L UNTIL N DO S := S + AB(J,K)*AB(I,K);
          FOR K := L UNTIL N DO AB(J,K) := AB(J,K) + S*E(K)
          END J
        END S;
      Y :=   ABS(Q(I)) + ABS(E(I));  IF Y >X THEN X := Y
      END I;

COMMENT: ACCUMULATION OF RIGHT-HAND TRANSFORMATIONS;
FOR I := N STEP -1 UNTIL 1 DO
    BEGIN
    IF G¬=0 THEN
      BEGIN
      H := AB(I,I+1)*G;
      FOR J := L UNTIL N DO AB(J,I) := AB(I,J)/H;
      FOR J := L UNTIL N DO
        BEGIN S := 0;
        FOR K := L UNTIL N DO S := S + AB(I,K)*AB(K,J);
        FOR K := L UNTIL N DO AB(K,J) := AB(K,J) + S*AB(K,I)
        END J
      END G;
    FOR J := L UNTIL N DO AB(I,J) := AB(J,I) := 0;
    AB(I,I) := 1;  G := E(I);  L := I
    END I;

COMMENT: DIAGONALIZATION OF THE BIDIAGONAL FORM;
EPS := EPS*X;
FOR K := N STEP -1 UNTIL 1 DO
    BEGIN KT := 0;
    TESTFSPLITTING:
    KT := KT + 1;  IF KT > 30 THEN
      BEGIN E(K) := OL;
      WRITE ("QR FAILED")
      END;
    FOR L2 := K STEP -1 UNTIL 1 DO
      BEGIN
      L := L2;
      IF ABS(E(L))<=EPS THEN GOTO TESTFCONVERGENCE;
      IF ABS(Q(L-1))<=EPS THEN GOTO CANCELLATION
      END L2;

    COMMENT: CANCELLATION OF E(L) IF L>1;
    CANCELLATION:
    C := 0;  S := 1;
    FOR I := L UNTIL K DO
      BEGIN
      F := S*E(I);  E(I) := C*E(I);
      IF ABS(F)<=EPS THEN GOTO TESTFCONVERGENCE;
      G := Q(I);  Q(I) := H := IF ABS(F) < ABS(G) THEN
      ABS(G)*LONGSQRT(1 + (F/G)**2) ELSE IF F ¬= 0 THEN
      ABS(F)*LONGSQRT(1 + (G/F)**2) ELSE 0;
```

```
        IF H = 0 THEN G := H := 1;
        COMMENT: THE ABOVE REPLACES Q(I):=H:=LONGSQRT(G*G+F*F)
                    WHICH MAY GIVE INCORRECT RESULTS IF THE
                    SQUARES UNDERFLOW OR IF F = G = 0;
        C := G/H; S := -F/H
        END I;

    TESTFCONVERGENCE:
    Z := Q(K); IF L=K THEN GOTO CONVERGENCE;

    COMMENT: SHIFT FROM BOTTOM 2*2 MINOR;
    X := Q(L); Y := Q(K-1); G := E(K-1); H := E(K);
    F := ((Y-Z)*(Y+Z) + (G-H)*(G+H))/(2*H*Y);
    G := LONGSQRT(F*F+1);
    F := ((X-Z)*(X+Z)+H*(Y/(IF F<0 THEN F-G ELSE F+G)-H))/X;

    COMMENT: NEXT QR TRANSFORMATION;
    C := S := 1;
    FOR I := L+1 UNTIL K DO
      BEGIN
      G := E(I); Y := Q(I); H := S*G; G := G*C;
      E(I-1) := Z := IF ABS(F) < ABS(H) THEN
      ABS(H)*LONGSQRT(1 + (F/H)**2) ELSE IF F ¬= 0 THEN
      ABS(F)*LONGSQRT(1 + (H/F)**2) ELSE 0;
      IF Z = 0 THEN Z := F := 1;
      C := F/Z; S := H/Z;
      F := X*C + G*S; G := -X*S +G*C; H := Y*S;
      Y := Y*C;
      FOR J := 1 UNTIL N DO
        BEGIN
        X := AB(J,I-1); Z := AB(J,I);
        AB(J,I-1) := X*C + Z*S; AB(J,I) := -X*S + Z*C
        END J;
      Q(I-1) := Z := IF ABS(F) < ABS(H) THEN ABS(H)*
      LONGSQRT(1 + (F/H)**2) ELSE IF F ¬= 0 THEN
      ABS(F)*LONGSQRT(1 + (H/F)**2) ELSE 0;
      IF Z = 0 THEN Z := F := 1;
      C := F/Z; S := H/Z;
      F := C*G + S*Y; X := -S*G + C*Y
      END I;
    E(L) := 0; E(K) := F; Q(K) := X;
    GO TO TESTFSPLITTING;

    CONVERGENCE:
    IF Z<0 THEN
      BEGIN COMMENT: Q(K) IS MADE NON-NEG;
      Q(K) := -Z;
      FOR J := 1 UNTIL N DO AB(J,K) := -AB(J,K)
      END Z
    END K
  END MINFIT;

PROCEDURE SORT;
  BEGIN COMMENT:   SORTS THE ELEMENTS OF D AND CORRESPONDING
                    COLUMNS OF V INTO DESCENDING ORDER;
  INTEGER K;
  LONG REAL S;
  FOR I := 1 UNTIL N - 1 DO
    BEGIN K := I; S := D(I); FOR J := I + 1 UNTIL N DO
    IF D(J) > S THEN
      BEGIN K := J; S := D(J) END;
    IF K > I THEN
      BEGIN D(K) := D(I); D(I) := S; FOR J := 1 UNTIL N DO
        BEGIN S := V(J,I); V(J,I) := V(J,K); V(J,K) := S
```

```
             END
           END
         END
       END SORT;

  PROCEDURE PRINT;
     COMMENT:  THE VARIABLE FMIN IS GLOBAL, AND ESTIMATES THE
               VALUE OF F AT THE MINIMUM: USED ONLY FOR
               PRINTING LOG(FX - FMIN);
     IF PRIN > 0 THEN
     BEGIN INTEGER SVINT;  SVINT := INTFIELDSIZE;
     INTFIELDSIZE := 10;
     WRITE (NL, NF, FX);
     COMMENT:  IF THE NEXT TWO LINES ARE OMITTED THEN FMIN IS
               NOT REQUIRED;
     IF FX <= FMIN THEN WRITEON ("      UNDEFINED  ") ELSE
     WRITEON (ROUNDTOREAL (LONGLOG (FX - FMIN)));
     COMMENT:  "IOCONTROL(2)" MOVES TO THE NEXT LINE;
     IF N > 4 THEN IOCONTROL(2);
     IF (N <= 4) OR (PRIN > 2) THEN
     FOR I := 1 UNTIL N DO WRITEON(ROUNDTOREAL(X(I)));
     IOCONTROL(2);  INTFIELDSIZE := SVINT
     END PRINT;

  PROCEDURE MATPRINT (STRING(80) VALUE S;  LONG REAL ARRAY
     V(*,*);  INTEGER VALUE M, N);
     BEGIN COMMENT:  PRINTS M X N MATRIX V COLUMN BY COLUMN;
     WRITE (S);
     FOR K := 1 UNTIL (N + 7) DIV 8 DO
       BEGIN FOR I := 1 UNTIL M DO
         BEGIN IOCONTROL(2);
         FOR J := 8*K - 7 UNTIL (IF N < (8*K) THEN N ELSE 8*K)
         DO WRITEON (ROUNDTOREAL (V (I,J)))
         END;
       WRITE (" ");  IOCONTROL(2)
       END
     END MATPRINT;

  PROCEDURE VECPRINT (STRING(32) VALUE S;  LONG REAL ARRAY V(*);
     INTEGER VALUE N);
     BEGIN COMMENT:  PRINTS THE HEADING S AND N-VECTOR V;
     WRITE(S);
     FOR I := 1 UNTIL N DO WRITEON(ROUNDTOREAL(V(I)))
     END VECPRINT;

  PROCEDURE MIN (INTEGER VALUE J, NITS;  LONG REAL VALUE
     RESULT D2, X1;  LONG REAL VALUE F1;  BOOLEAN VALUE FK);
     BEGIN COMMENT:
                    MINIMIZES F FROM X IN THE DIRECTION V(*,J)
                    UNLESS J<1, WHEN A QUADRATIC SEARCH IS DONE
                    IN THE PLANE DEFINED BY Q0, Q1 AND X.
                    D2 AN APPROXIMATION TO HALF F'' (OR ZERO),
                    X1 AN ESTIMATE OF DISTANCE TO MINIMUM,
                    RETURNED AS THE DISTANCE FOUND.
                    IF FK = TRUE THEN F1 IS FLIN(X1), OTHERWISE
                    X1 AND F1 ARE IGNORED ON ENTRY UNLESS FINAL
                    FX > F1.  NITS CONTROLS THE NUMBER OF TIMES
                    AN ATTEMPT IS MADE TO HALVE THE INTERVAL.
          SIDE EFFECTS: USES AND ALTERS X, FX, NF, NL.
                    IF J < 1 USES VARIABLES Q... .
                    USES H, N, T, M2, M4, LDT, DMIN, MACHEPS;

     LONG REAL PROCEDURE FLIN (LONG REAL VALUE L);
        COMMENT:  THE FUNCTION OF ONE VARIABLE L WHICH IS
                  MINIMIZED BY PROCEDURE MIN;
        BEGIN LONG REAL ARRAY T(1::N);
```

```
    IF J > 0 THEN
      BEGIN COMMENT:  LINEAR SEARCH;
      FOR I := 1 UNTIL N DO T(I) := X(I) + L*V(I,J)
      END
    ELSE
      BEGIN COMMENT:  SEARCH ALONG A PARABOLIC SPACE-CURVE;
      QA := L*(L - QD1)/(QD0*(QC0 + QD1));
      QB := (L + QD0)*(QD1 - L)/(QD0*QD1);
      QC := L*(L + QD0)/(QD1*(QC0 + QD1));
      FOR I := 1 UNTIL N DO T(I) := QA*Q0(I)+QB*X(I)+QC*Q1(I)
      END;
    COMMENT:  INCREMENT FUNCTION EVALUATION COUNTER;
    NF := NF + 1;
    F(T, N)
    END FLIN;

INTEGER K;  BOOLEAN DZ;
LONG REAL X2, XM, F0, F2, FM, D1, T2, S, SF1, SX1;
SF1 := F1;   SX1 := X1;
K := 0;   XM := 0;   F0 := FM := FX;   DZ := (D2 < MACHEPS);
COMMENT:  FIND STEP SIZE;
S := 0;   FOR I := 1 UNTIL N DO S := S + X(I)**2;
S := LONGSQRT(S);
T2:= M4*LONGSQRT(ABS(FX)/(IF DZ THEN DMIN ELSE D2)
     + S*LDT) + M2*LDT;
S := M4*S + T;
IF DZ AND (T2 > S) THEN T2 := S;
IF T2 < SMALL THEN T2 := SMALL;
IF T2 > (0.01*H) THEN T2 := 0.01*H;
IF FK AND (F1 <= FM) THEN XM := X1;   FM := F1 END;
IF ¬FK OR (ABS(X1) < T2) THEN
  BEGIN X1 := IF X1 >= 0L THEN T2 ELSE -T2;
  F1 := FLIN(X1)
  END;
IF F1 <= FM THEN BEGIN XM := X1;   FM := F1 END;
L0:  IF DZ THEN
  BEGIN COMMENT:   EVALUATE FLIN AT ANOTHER POINT AND
                   ESTIMATE THE SECOND DERIVATIVE;
  X2 := IF F0 < F1 THEN -X1 ELSE 2*X1;   F2 := FLIN(X2);
  IF F2 <= FM THEN BEGIN XM := X2;   FM := F2 END;
  D2 := (X2*(F1 - F0) - X1*(F2 - F0))/(X1*X2*(X1 - X2))
  END;
COMMENT:  ESTIMATE FIRST DERIVATIVE AT 0;
D1 := (F1 - F0)/X1 - X1*D2;   DZ := TRUE;
COMMENT:  PREDICT MINIMUM;
X2 := IF D2 <= SMALL THEN (IF D1 < 0 THEN H ELSE -H) ELSE
     -0.5L*D1/D2;
IF ABS(X2) > H THEN X2 := IF X2 > 0 THEN H ELSE -H;
COMMENT:  EVALUATE F AT THE PREDICTED MINIMUM;
L1:  F2 := FLIN(X2);
IF (K < NITS) AND (F2 > F0) THEN
  BEGIN COMMENT:  NO SUCCESS SO TRY AGAIN;   K := K + 1;
  IF (F0 < F1) AND ((X1*X2) > 0) THEN GO TO L0;
  X2 := 0.5L*X2;   GO TO L1
  END;
COMMENT:  INCREMENT ONE-DIMENSIONAL SEARCH COUNTER;
NL := NL + 1;
IF F2 > FM THEN X2 := XM ELSE FM := F2;
COMMENT:  GET NEW ESTIMATE OF SECOND DERIVATIVE;
D2 := IF ABS(X2*(X2 - X1)) > SMALL THEN
     (X2*(F1 - F0) - X1*(FM - F0))/(X1*X2*(X1 - X2))
     ELSE IF K > 0 THEN 0 ELSE D2;
IF D2 <= SMALL THEN D2 := SMALL;
X1 := X2;   FX := FM;
IF SF1 < FX THEN BEGIN FX := SF1;   X1 := SX1 END;
```

```
      COMMENT:   UPDATE X FOR LINEAR SEARCH BUT NOT FOR PARABOLIC
                 PARABOLIC SEARCH;
      IF J > 0 THEN FOR I := 1 UNTIL N DO X(I) := X(I) + X1*V(I,J)
    END MIN;

PROCEDURE QUAD;
   BEGIN COMMENT:   LOOKS FOR THE MINIMUM ALONG A CURVE
                    DEFINED BY Q0, Q1 AND X;
   LONG REAL L, S;
   S := FX;   FX := QF1;   QF1 := S;   QD1 := 0;
   FOR I := 1 UNTIL N DO
     BEGIN S := X(I);   X(I) := L := Q1(I);   Q1(I) := S;
     QD1 := QD1 + (S - L)**2
     END;
   L := QD1 := LONGSQRT(QD1);   S := 0;
   IF (QD0 > 0) AND (QD1 > 0) AND (NL >= (3*N*N)) THEN
     BEGIN MIN (0, 2, S, L, QF1, TRUE);
     QA := L*(L - QD1)/(QD0*(QD0 + QD1));
     QB := (L + QD0)*(QD1 - L)/(QD0*QD1);
     QC := L*(L + QD0)/(QD1*(QD0 + QD1))
     END
   ELSE BEGIN FX := QF1;   QA := QB := 0;   QC := 1 END;
   QD0 := QD1;   FOR I := 1 UNTIL N DO
     BEGIN S := Q0(I);   Q0(I) := X(I);
     X(I) := QA*S + QB*X(I) + QC*Q1(I)
     END
   END QUAD;


BOOLEAN ILLC;
INTEGER NL, NF, KL, KT, KTM;
LONG REAL S, SL, DN, DMIN, FX, F1, LDS, LDT, SF, DF,
QF1, QD0, QD1, QA, QB, QC,
M2, M4, SMALL, VSMALL, LARGE, VLARGE, SCBD, LDFAC, T2;
LONG REAL ARRAY D, Y, Z, Q0, Q1 (1::N);
LONG REAL ARRAY V (1::N, 1::N);

COMMENT:   INITIALIZATION;
COMMENT:   MACHINE DEPENDENT NUMBERS;
SMALL := MACHEPS**2;   VSMALL := SMALL**2;
LARGE := 1L/SMALL;       VLARGE := 1L/VSMALL;
M2 := LONGSQRT(MACHEPS);   M4 := LONGSQRT(M2);

COMMENT:   HEURISTIC NUMBERS
           ******************

   IF AXES MAY BE BADLY SCALED (WHICH IS TO BE AVOIDED IF
POSSIBLE) THEN SET SCBD := 10, OTHERWISE 1.
   IF THE PROBLEM IS KNOWN TO BE ILLCONDITIONED SET
ILLC := TRUE, OTHERWISE FALSE.
   KTM+1 IS THE NUMBER OF ITERATIONS WITHOUT IMPROVEMENT BEFORE
THE ALGORITHM TERMINATES (SEE SECTION 6).   KTM = 4 IS VERY
CAUTIOUS:   USUALLY KTM = 1 IS SATISFACTORY;

SCBD := 1;   ILLC := FALSE;   KTM := 1;

LDFAC := IF ILLC THEN 0.1 ELSE 0.01;
KT := NL := 0;   NF := 1;   QF1 := FX := F(X,N);
T := T2 := SMALL + ABS(T);   DMIN := SMALL;
IF H < (100*T) THEN H := 100*T;   LDT := H;
FOR I := 1 UNTIL N DO FOR J := 1 UNTIL N DO
V(I,J) := IF I = J THEN 1L ELSE 0L;
D(1) := QD0 := 0;   FOR I := 1 UNTIL N DO Q1(I) := X(I);
PRINT;

COMMENT: MAIN LOOP;
L0:   SF := D(1);   D(1) := S := 0;
```

```
COMMENT:  MINIMIZE ALONG FIRST DIRECTION;
MIN (1, 2, D(1), S, FX, FALSE);
IF S <= 0 THEN FOR I := 1 UNTIL N DO V(I,1) := -V(I,1);
IF (SF <= (0.9*D(1))) OR ((0.9*SF) >= D(1)) THEN
FOR I := 2 UNTIL N DO D(I) := 0;
FOR K := 2 UNTIL N DO
  BEGIN FOR I := 1 UNTIL N DO Y(I) := X(I);  SF := FX;
  ILLC := ILLC OR (KT > 0);
  L1:  KL := K;  DF := 0;  IF ILLC THEN
    BEGIN COMMENT: RANDOM STEP TO GET OFF RESOLUTION VALLEY;
    FOR I := 1 UNTIL N DO
      BEGIN S := Z(I) := (0.1*LDT + T2*10**KT)*(RANDOM-0.5L);
      COMMENT:   PRAXIS ASSUMES THAT RANDOM RETURNS A RANDOM
                 NUMBER UNIFORMLY DISTRIBUTED IN (0, 1) AND
                 THAT ANY INITIALIZATION OF THE RANDOM NUMBER
                 GENERATOR HAS ALREADY BEEN DONE;
      FOR J := 1 UNTIL N DO X(J) := X(J) + S*V(J,I)
      END;
    FX := F(X, N);  NF := NF + 1
    END;
  FOR K2 := K UNTIL N DO
    BEGIN SL := FX;  S := 0;
    COMMENT:  MINIMIZE ALONG "NON-CONJUGATE" DIRECTIONS;
    MIN (K2, 2, D(K2), S, FX, FALSE);
    S := IF ILLC THEN D(K2)*(S + Z(K2))**2 ELSE SL - FX;
    IF DF < S THEN
      BEGIN DF := S;  KL := K2
      END
    END;
  IF ¬ILLC AND (DF < ABS(100*MACHEPS*FX)) THEN
    BEGIN COMMENT:  NO SUCCESS  ILLC = FALSE SO TRY ONCE
                    WITH ILLC = TRUE;
    ILLC := TRUE;  GO TO L1
    END;
  IF (K = 2) AND (PRIN > 1) THEN VECPRINT ("NEW D", D, N);
  FOR K2 := 1 UNTIL K - 1 DO
    BEGIN COMMENT: MINIMIZE ALONG "CONJUGATE" DIRECTIONS;
    S := 0;  MIN (K2, 2, D(K2), S, FX, FALSE)
    END;
  F1 := FX;  FX := SF;  LDS := 0;
  FOR I := 1 UNTIL N DO
    BEGIN SL := X(I);  X(I) := Y(I);  SL := Y(I) := SL - Y(I);
    LDS := LDS + SL*SL
    END;
  LDS := LONGSQRT(LDS);  IF LDS > SMALL THEN
    BEGIN COMMENT:   THROW AWAY DIRECTION KL AND MINIMIZE
                     ALONG THE NEW "CONJUGATE" DIRECTION;
    FOR I := KL - 1 STEP -1 UNTIL K DO
      BEGIN FOR J := 1 UNTIL N DO V(J,I + 1) := V(J,I);
      D(I + 1) := D(I)
      END;
    D(K) := 0;    FOR I := 1 UNTIL N DO V(I,K) := Y(I)/LDS;
    MIN (K, 4, D(K), LDS, F1, TRUE);
    IF LDS <= 0 THEN
      BEGIN LDS := -LDS;
      FOR I := 1 UNTIL N DO V(I,K) := -V(I,K)
      END
    END;
  LDT := LDFAC*LDT;  IF LDT < LDS THEN LDT := LDS;
  PRINT;
  T2 := 0;  FOR I := 1 UNTIL N DO T2 := T2 + X(I)**2;
  T2 := M2*LONGSQRT(T2) + T;
  COMMENT:  SEE IF STEP LENGTH EXCEEDS HALF THE TOLERANCE;
  KT := IF LDT > (0.5*T2) THEN 0 ELSE KT + 1;
  IF KT > KTM THEN GO TO L2
  END;
```

```
COMMENT:    TRY QUADRATIC EXTRAPOLATION IN CASE WE ARE STUCK
            IN A CURVED VALLEY;
QUAD;
DN := 0;  FOR I := 1 UNTIL N DO
  BEGIN D(I) := 1/LONGSQRT(D(I));
  IF DN < D(I) THEN DN := D(I)
  END;
IF PRIN > 3 THEN MATPRINT ("NEW DIRECTIONS", V, N, N);
FOR J := 1 UNTIL N DO
  BEGIN S := D(J)/DN;
  FOR I := 1 UNTIL N DO V(I,J) := S*V(I,J)
  END;
IF SCBD > 1 THEN
  BEGIN COMMENT:   SCALE AXES TO TRY TO RECUCE CONDITION
                   NUMBER;
  S := VLARGE;  FOR I := 1 UNTIL N DO
    BEGIN SL := 0;  FOR J := 1 UNTIL N DO SL := SL+V(I,J)**2;
    Z(I) := LONGSQRT(SL);
    IF Z(I) < M4 THEN Z(I) := M4;  IF S > Z(I) THEN S := Z(I)
    END;
  FOR I := 1 UNTIL N DO
    BEGIN SL := S/Z(I);  Z(I) := 1/SL;  IF Z(I) > SCBD THEN
      BEGIN SL := 1/SCBD;  Z(I) := SCBD
      END;
  FOR J := 1 UNTIL N DO V(I,J) := SL* V(I,J) END END;
COMMENT:  TRANSPOSE V FOR MINFIT;
FOR I := 2 UNTIL N DO FOR J := 1 UNTIL I - 1 DO
  BEGIN S := V(I,J);  V(I,J) := V(J,I);   V(J,I) := S END;
COMMENT:   FIND THE SINGULAR VALUE DECOMPOSITION OF V.   THIS
           GIVES THE EIGENVALUES AND PRINCIPAL AXES OF THE
           APPROXIMATING QUADRATIC FORM WITHOUT SQUARING THE
           CONDITION NUMBER;
MINFIT (N, MACHEPS, VSMALL, V, D);
IF SCBD > 1 THEN
  BEGIN COMMENT:  UNSCALING;  FOR I := 1 UNTIL N DO
    BEGIN S := Z(I);
    FOR J := 1 UNTIL N DO V(I,J) := S*V(I,J)
    END;
  FOR I := 1 UNTIL N DO
    BEGIN S := 0;  FOR J := 1 UNTIL N DO S := S + V(J,I)**2;
    S := LONGSQRT(S);   D(I) := S*D(I);  S := 1/S;
    FOR J := 1 UNTIL N DO V(J,I) := S*V(J,I)
    END
  END;
FOR I := 1 UNTIL N DO
  BEGIN D(I) := IF (DN*D(I)) > LARGE THEN VSMALL ELSE
  IF (DN*D(I)) < SMALL THEN VLARGE ELSE (DN*D(I))**(-2)
  END;
COMMENT:  SORT NEW EIGENVALUES AND EIGENVECTORS;
SORT;
DMIN := D(N);  IF DMIN < SMALL THEN DMIN := SMALL;
ILLC := (M2*D(1)) > DMIN;
IF (PRIN > 1) AND (SCBD > 1) THEN
VECPRINT ("SCALE FACTORS", Z, N);
IF PRIN > 1 THEN VECPRINT ("EIGENVALUES OF A", D, N);
IF PRIN > 3 THEN MATPRINT ("EIGENVECTORS OF A", V, N, N);
COMMENT:  GO BACK TO MAIN LOOP;
GO TO LO;
L2:  IF PRIN > 0 THEN VECPRINT ("X IS", X, N);
FX
END PRAXIS;
```

COMMENT: RANDOM NUMBER GENERATOR
                           ****

        PROCEDURE RANDOM RETURNS A LONG REAL RANDOM NUMBER UNIFORMLY

```
DISTRIBUTED IN (0,1) (INCLUDING 0 BUT NOT 1).
    RANINIT(R) WITH R ANY INTEGER MUST BE CALLED FOR
INITIALIZATION BEFORE THE FIRST CALL TO RANDOM, AND THE
DECLARATIONS OF RAN1, RAN2 AND RAN3 MUST BE GLOBAL.
    THE ALGORITHM RETURNS X(N)/2**56, WHERE
                    X(N) = X(N-1) + X(N-127)   (MOD 2**56).
SINCE 1 + X + X**127 IS PRIMITIVE (MOD 2), THE PERIOD IS AT
LEAST 2**127 - 1 > 10**38.    SEE KNUTH (1969), PP. 26, 34, 464.
    X(N) IS STORED IN A LONG REAL WORD AS
RAN3 = X(N)/2**56 - 1/2, AND ALL FLOATING POINT ARITHMETIC
IS EXACT;

LONG REAL RAN1;   INTEGER RAN2;   LONG REAL ARRAY RAN3 (0::126);

PROCEDURE RANINIT (INTEGER VALUE R);
  BEGIN R := ABS(R) REM 8190 + 1;
  RAN2 := 127;   WHILE RAN2 > 0 DO
    BEGIN RAN2 := RAN2 - 1;   RAN1 := -2L**55;
    FOR I := 1 UNTIL 7 DO
      BEGIN R := (1756*R) REM 8191;
      RAN1 := (RAN1 + (R DIV 32))*(1/256);
      END;
    RAN3 (RAN2) := RAN1
    END
  END RANINIT;

LONG REAL PROCEDURE RANDOM;
  BEGIN RAN2 := IF RAN2 = 0 THEN 126 ELSE RAN2 - 1;
  RAN1 := RAN1 + RAN3 (RAN2);
  RAN3 (RAN2) := RAN1 := IF RAN1 < 0L THEN RAN1 + 0.5L
                          ELSE RAN1 - 0.5L;
  RAN1 + 0.5L
  END RANDOM;

COMMENT:   TEST FUNCTIONS
           **************;

LONG REAL PROCEDURE ROS (LONG REAL ARRAY X(*); INTEGER VALUE N);
  COMMENT:   SEE ROSENBROCK (1960);
  100L*((X(2) - X(1)**2)**2) + (1L - X(1))**2;

LONG REAL PROCEDURE SING(LONG REAL ARRAY X(*);INTEGER VALUE N);
  COMMENT:   SEE POWELL (1962);
  (X(1) + 10L*X(2))**2 + 5L*(X(3)-X(4))**2 + (X(2)-2L*X(3))**4
  + 10L*(X(1) - X(4))**4;

LONG REAL PROCEDURE HELIX(LONG REAL ARRAY X(*);INTEGER VALUE N);
  COMMENT:   SEE FLETCHER & POWELL (1963);
  BEGIN LONG REAL R, T;
  R := LONGSQRT (X(1)**2 + X(2)**2);
  T := IF X(1) = 0 THEN 0.25L ELSE LONGARCTAN (X(2)/X(1))/(2L*
       3.14159265358979L);
  IF X(1) < 0 THEN T := T + 0.5L;
  100L*((X(3) - 10L*T)**2 + (R - 1L)**2) + X(3)**2
  END HELIX;

LONG REAL PROCEDURE CUBE(LONG REAL ARRAY X(*);INTEGER VALUE N);
  COMMENT:   SEE LEON (1966);
  100L*(X(2) - X(1)**3)**2 + (1L - X(1))**2;

LONG REAL PROCEDURE BEALE(LONG REAL ARRAY X(*);INTEGER VALUE N);
  COMMENT:   SEE BEALE (1958);
  (1.5L - X(1)*(1L - X(2)))**2 +
  (2.25L - X(1)*(1L - X(2)**2))**2 +
  (2.625L - X(1)*(1L - X(2)**3))**2;

LONG REAL PROCEDURE WATSON (LONG REAL ARRAY X(*);
```

```
   INTEGER VALUE N);
   COMMENT:  SEE KOWALIK & OSBORNE (1968);
   BEGIN LONG REAL S, T, U, Y;
   S := X(1)**2 + (X(2) - X(1)**2 - 1L)**2;
   FOR I := 2 UNTIL 30 DO
      BEGIN Y := (I - 1)/29;  T := X(N);
      FOR J := N - 1 STEP -1 UNTIL 1 DO T := X(J) + Y*T;
      U := (N - 1)*X(N);
      FOR J := N - 1 STEP -1 UNTIL 2 DO U := (J - 1)*X(J) + Y*U;
      S := S + (U - T*T - 1L)**2
      END;
   S
   END WATSON;

LONG REAL PROCEDURE CHEBYQUAD (LONG REAL ARRAY X(*);
   INTEGER VALUE N);
   COMMENT:  SEE FLETCHER (1965);
   BEGIN
   LONG REAL F, DELTA, TPLUS;
   BOOLEAN EVEN;
   LONG REAL ARRAY Y, TI, TMINUS (1::N);

   DELTA := OL;
   FOR J := 1 UNTIL N DO
      BEGIN Y(J) := 2L*X(J) - 1L;
      DELTA := DELTA + Y(J);
      TI(J) := Y(J);  TMINUS(J) := 1L
      END;
   F := DELTA**2;  EVEN := FALSE;
   FOR I := 2 UNTIL N DO
      BEGIN EVEN := ¬EVEN;  DELTA := OL;
      FOR J := 1 UNTIL N DO
         BEGIN TPLUS := 2L*Y(J)*TI(J) - TMINUS(J);
         DELTA := DELTA + TPLUS;
         TMINUS(J) := TI(J);
         TI(J) := TPLUS
         END;
      DELTA := DELTA/N - (IF EVEN THEN 1/(1 - I*I) ELSE 0);
      F := F + DELTA**2
      END;
   F
   END CHEBYQUAD;

LONG REAL PROCEDURE POWELL (LONG REAL ARRAY X(*);
   INTEGER VALUE N);
   COMMENT:  SEE POWELL (1964);
   3L - 1L/(1L + (X(1) - X(2))**2) -
   LONGSIN(0.5L*3.14159265358979L*X(2)* X(3))-(IF X(2) = 0 THEN
   OL ELSE LONGEXP(-((X(1)+X(3))/X(2) - 2L)**2));

LONG REAL PROCEDURE WOOD(LONG REAL ARRAY X(*);INTEGER VALUE N);
   COMMENT:  SEE MCCORMICK & PEARSON (1969) OR COLVILLE (1968);
   100L*(X(2) - X(1)**2)**2 + (1L - X(1))**2 + 90L*(X(4) -
   X(3)**2)**2 + (1L - X(3))**2 + 10.1L*((X(2) - 1L)**2 + (X(4)
   - 1L)**2) + 19.8L*(X(2) - 1L)*(X(4) - 1L);

LONG REAL PROCEDURE HILBERT (LONG REAL ARRAY X(*);
   INTEGER VALUE N);
   COMMENT:  COMPUTES XT.A.X, WHERE A IS THE N BY N HILBERT
             MATRIX, SEE GREGORY & KARNEY (1969), PP. 33, 66;
   BEGIN LONG REAL S, T;
   S := OL;  FOR I := 1 UNTIL N DO
      BEGIN T := OL;  FOR J := 1 UNTIL N DO
      T := T + X(J)/(I + J - 1);
```

```
      S := S + T*X(I)
      END;
    S
    END HILBERT;

LONG REAL PROCEDURE TRIDIAG (LONG REAL ARRAY X(*);
  INTEGER VALUE N);
  COMMENT:  COMPUTES XT.A.X - 2E1T.X, WHERE N > 1,

                      ( 1 -1  0  0 ... 0)
                      (-1  2 -1  0 ... 0)
                      ( 0 -1  2 -1 ... 0)
             A  =      (.................)
                      ( 0  ...   -1  2 -1)
                      ( 0  ...    0 -1  2),

          AND E1T = (1, 0, ... , 0).

          SEE GREGORY & KARNEY (1969), PP. 41, 74;
  BEGIN LONG REAL S;
  S := X(1)*(X(1) - X(2));
  FOR I := 2 UNTIL N - 1 DO
  S := S + X(I)*((X(I) - X(I - 1)) + (X(I) - X(I + 1)));
  S + X(N)*(2*X(N) - X(N - 1)) - 2*X(1)
  END TRIDIAG;

LONG REAL PROCEDURE BOX (LONG REAL ARRAY X(*);INTEGER VALUE N);
  COMMENT:  SEE BOX (1966) OR BROWN & DENNIS (1970);
  BEGIN LONG REAL P, S;
  S := 0;  FOR I := 1 UNTIL 10 DO
    BEGIN P := -I/10;
    S := S + ((LONGEXP(P*X(1))) - (IF (P*X(2)) < (-40) THEN 0
        ELSE LONGEXP(P*X(2)))) -
        X(3)*(LONGEXP(P) - LONGEXP(10*P)))**2
    END;
  S
  END BOX;

COMMENT:  GENERAL TESTING PROCEDURE
          ***********************;

PROCEDURE TEST (STRING (80) VALUE S;  LONG REAL VALUE H;
                LONG REAL PROCEDURE F;  INTEGER VALUE N);
  BEGIN LONG REAL FMIN;  INTEGER TIM;
  WRITE(" ");  WRITE(" ");  WRITE(S);
  WRITE("N =", N, "  H =", ROUNDTOREAL(H));  WRITE(" ");
  COMMENT:  INITIALIZE RANDOM NUMBER GENERATOR;  RANINIT(4);
  COMMENT:  TIME(2) RETURNS CLOCK TIME IN UNITS OF 26 MICROSEC;
  TIM := TIME(2);.
  FMIN := PRAXIS (1'-5, 16**(-13), H, N, 1, X, F, RANDOM);
  WRITE ("TIME (MILLISEC) =", ROUND((TIME(2) - TIM)/38.4));
  WRITE(" ")
  END TEST;

COMMENT:  TESTING PROGRAM
          **************;

LONG REAL FMIN, LAM;
COMMENT:  INCREASE DIMENSIONS FOR N > 20;
LONG REAL ARRAY X(1::20);
COMMENT:  INTFIELDSIZE CONTROLS THE OUTPUT FORMAT OF INTEGERS;
INTFIELDSIZE := 7;

X(1) := -1.2L;  X(2) := 1L;  FMIN := 0;
TEST ("ROSENBROCK'S FUNCTION WITH A PARABOLIC VALLEY",1,ROS,2);
```

```
      X(1) := X(2) := 3;
      TEST ("ROSENBROCK'S FUNCTION ", 3, ROS, 2);

      X(1) := X(2) := 8;
      TEST ("ROSENBROCK'S FUNCTION", 12, ROS, 2);

      X(1) := -1;  X(2) := X(3) := 0;
      TEST ("HELIX", 1, HELIX, 3);

      X(1) := -1.2L;  X(2) := -1;
      TEST ("CUBE", 1, CUBE, 2);

      X(1) := X(2) := 0.1L;
      TEST ("BEALE", 1, BEALE, 2);

      X(1) := 0;  X(2) := 1;  X(3) := 2;
      TEST ("POWELL", 1, POWELL, 3);

      FMIN := 0;  X(1) := 0;  X(2) := 10;  X(3) := 20;
      TEST ("BOX", 20, BOX, 3);

      X(1) := 3L;  X(2) := -1L;  X(3) := 0L;  X(4) := 1L;
      TEST ("POWELL'S FUNCTION WITH A SINGULAR JACOBIAN",1,SING,4);

      FMIN := 0;  X(1) := X(3) := -3;  X(2) := X(4) := -1;
      TEST ("WOOD", 10, WOOD, 4);

      FOR N := 2 STEP 2 UNTIL 8 DO
        BEGIN FOR I := 1 UNTIL N DO X(I) := I/(N + 1);
        FMIN := IF N < 8 THEN 0L ELSE 0.0035168737256779L;
        TEST ("CHEBYQUAD", 0.1, CHEBYQUAD, N)
        END;
      FOR N := 6 STEP 3 UNTIL 9 DO
        BEGIN FOR I := 1 UNTIL N DO X(I) := 0;
        FMIN := IF N = 6 THEN 0.00228767005355L ELSE
                IF N = 9 THEN 1.399760138098'-6L ELSE 0L;
        TEST ("WATSON", 1, WATSON, N)
        END;

      FOR N := 4, 6, 8, 10, 12, 16, 20 DO
        BEGIN FOR I := 1 UNTIL N DO X(I) := 0L;  FMIN := -N;
        TEST ("TRIDIAG", 2*N, TRIDIAG, N)
        END;

      FMIN := 0;  FOR N := 2 STEP 2 UNTIL 12 DO
        BEGIN FOR I := 1 UNTIL N DO X(I) := 1;
        TEST ("HILBERT", 10, HILBERT, N)
        END
      END.
```

BIBLIOGRAPHY

This bibliography contains references relevant to the minimization of nonlinear functions. There is no attempt at completeness, but many recent references on unconstrained minimization have been included. There are also some references dealing with constrained problems, with methods for converting constrained problems to unconstrained problems, and with methods for solving nonlinear equations. For a brief survey, see Section 7.1. References on linear and quadratic programming have generally been excluded, and we have not attempted to duplicate the large bibliographies in Jacoby, Kowalik, and Pizzo (1971); Künzi and Oettli (1970); Lawson (1968); and Ortega and Rheinboldt (1970).

In lieu of annotations, the chapter and section numbers of references to each entry are given in parentheses after the entry.

References "to appear" have arbitrarily been assigned the year 1971.

Abadie, J. (ed.), 1970, *Nonlinear and integer programming*, North-Holland, Amsterdam. (7.1)

Akaike, H., 1959, On a successive transformation of probability distribution and its application to the analysis of the optimum gradient method, *Ann. Inst. Statist. Math. of Tokyo* 11, 1–16. (7.1)

Akilov, G.P., see Kantorovich and Akilov (1959).

Allran, R. R., and Johnsen, S. E. J., 1970, An algorithm for solving nonlinear programming problems subject to nonlinear inequality constraints, *Comp. J.* 13, 2, 171–177. (7.1)

Andrews, A. M., 1969, The calculation of orthogonal vectors, *Comp. J.* 12, 411. (7.5)

Armijo, L., 1966, Minimization of functions having Lipschitz continuous first partial derivatives, *Pacific J. Math.* 16, 1–3. (1.2)

Avriel, M., and Wilde, D. J., 1966, Optimal search for a maximum with sequences of simultaneous function evaluations, *Mgmt. Sci.* 12, 722–731. (5.7)

Baer, R. M., 1962, Note on an extremum locating algorithm, *Comp. J.* 5, 193. (7.5)

Baker, C. T. H., 1970, The error in polynomial interpolation, *Numer. Math.* 15, 315–319. (2.4)

Balakrishnan, A. V., 1970, see *Symposium on optimization (Nice, June 1969)*, Springer-Verlag, Berlin. (7.1)

Bard, Y., 1968, On a numerical instability of Davidon-like methods, *Math. Comp.* 22, 665–666. (7.1)

Bard, Y., 1970, Comparison of gradient methods for the solution of nonlinear parameter estimation problems, *SIAM J. Numer. Anal.* 7, 157–186. (7.1)

Bard, Y., see Greenstadt (1970).

Barnes, J. P. G., 1965, An algorithm for solving nonlinear equations based on the secant method, *Comp. J.* 8, 66–72. (7.1)

Bartels, R. H., 1968, *A numerical investigation of the simplex method*, Report CS 104, Computer Sci. Dept., Stanford Univ. (7.1)

Bartels, R. H., and Golub, G. H., 1969, The simplex method of linear programming using LU decomposition, *Comm. ACM* 12, 266–268. (7.1)

Bartels, R. H., Golub, G. H., and Saunders, M. A., 1970, *Numerical techniques in mathematical programming*, Report CS 162, Computer Sci. Dept., Stanford Univ. (7.1)

Bauer, H., Becker, S., and Graham, S., 1968, *ALGOL W language description*, Report CS 89 (revised as CS 110 with E. Satterthwaite, 1969), Computer Sci. Dept., Stanford Univ. (4.4, 5.6, 6.6, 7.9)

Beale, E. M. L., 1958, *On an iterative method for finding a local minimum of a function of more than one variable*, Tech. Report No. 25, Statistical Techniques Research Group, Princeton Univ. (7.7, 7.9)

Beale, E. M. L., 1968, *Mathematical programming in practice*, Wiley, New York. (7.1)

Becker, S., see Bauer, Becker, and Graham (1968).

Beightler, C. S., see Wilde and Beightler (1967).

Bell, M., and Pike, M. C., 1966, Remark on algorithm 178(E4), Direct Search, *Comm. ACM* 9, 684. (7.1)

Bellman, R. E., 1957, *Dynamic programming*, Princeton Univ. Press, Princeton, New Jersey. (1.2)

Bellman, R. E., and Dreyfus, S. E., 1962, *Applied dynamic programming*, Princeton Univ. Press, Princeton, New Jersey. (1.2, 4.1)

Bennett, J. M., 1965, Triangular factors of modified matrices, *Numer. Math.* 7, 217–221. (7.1)

Berman, G., 1969, Lattice approximations to the minima of functions of several variables, *J. ACM* 16, 286–294. (7.1)

Björck, A., 1967a, Solving linear least squares problems by Gram-Schmidt orthogonalization, *BIT* 7, 1–21. (7.1)

Björck, A., 1967b, Iterative refinement of linear least squares solutions I, *BIT* 7, 257–278. (7.1)

Björck, A., 1968, Iterative refinement of linear least squares solutions II, *BIT* 8, 8–30. (7.1)

Boothroyd, J., 1965a, Algorithm 7, MINIX, *Comp. Bulletin* 9, 104. (5.3)

Boothroyd, J., 1965b, Certification of Algorithm 2, Fibonacci Search, *Comp. Bulletin* 9, 105. (5.3)

Bowdler, H., Martin, R. S., Reinsch, C., and Wilkinson, J. H., 1968, The QR and QL algorithms for symmetric matrices, *Numer. Math.* 11, 293–306. (7.4)

Box, G. E. P., 1957, Evolutionary operations: a method for increasing industrial productivity, *Appl. Stat.* 6, 3–23. (7.1)

Box, M. J., 1965, A new method for constrained optimization and a comparison with other methods, *Comp. J.* 8, 42–52. (7.1)

Box, M. J., 1966, A comparison of several current optimization methods, and the use of transformations in constrained problems, *Comp. J.* 9, 67–77. (7.1, 7.3, 7.7, 7.9)

Box, M. J., Davies, D., and Swann, W. H., 1969, *Non-linear optimization techniques*, ICI Monograph No. 5, Oliver and Boyd, London. (5.4, 5.5, 7.1)

Brent, R. P., 1971a, *A note on the Davidenko-Branin method for solving nonlinear equations*, Report RC 3506, IBM T. J. Watson Research Lab., Yorktown Heights, New York, to appear in *IBM Jour. Res. and Dev.* (7.1)

Brent, R. P., 1971b, On maximizing the efficiency of algorithms for solving systems of nonlinear equations, Report RC 3725, IBM, Yorktown Heights. (7.1)

Brent, R. P., 1971c, An efficient algorithm for unconstrained optimization without derivatives, to appear. (7.9)

Brent, R. P., 1971d, An algorithm with guaranteed convergence for finding a zero of a function, *Comp. J.* 14, 422–25.

Brown, K. M., and Conte, S. D., 1967, The solution of simultaneous nonlinear equations, *Proc. 22nd National Conference of the ACM*, Thompson Book Co., Washington, D. C., 111–114. (7.1)

Brown, K. M., and Dennis, J. E., 1968, On Newton-like iteration functions: general convergence theorems and a specific algorithm, *Numer. Math.* 12, 186–191. (7.1)

Brown, K. M., and Dennis, J. E., 1971a, On the second order convergence of Brown's method for solving simultaneous nonlinear equations, to appear. (7.1)

Brown, K. M., and Dennis, J. E., 1971b, Derivative-free analogues of the Levenberg-Marquardt and Gauss algorithms for nonlinear least squares approximation, to appear in *Numer. Math.* (7.1)

Broyden, C. G., 1965, A class of methods for solving nonlinear simultaneous equations, *Math. Comp.* 19, 577–593. (7.1)

Broyden, C. G., 1967, Quasi-Newton methods and their application to function minimization, *Math. Comp.* 21, 368–381. (7.1, 7.7)

Broyden, C. G., 1969, A new method of solving nonlinear simultaneous equations, *Comp. J.* 12, 94–99. (7.1)

Broyden, C. G., 1970a, The convergence of a class of double-rank minimization algorithms, Parts I and II, *J. Inst. Maths. Apps.* 6, 76–90 and 222–231. (7.1)

Broyden, C. G., 1970b, The convergence of single-rank quasi-Newton methods, *Math. Comp.* 24, 365–382. (7.1)

Buehler, R. J., see Shah, Buehler, and Kempthorne (1964).

Businger, P., and Golub, G. H., 1965, Linear least squares solutions by Householder transformations, *Numer. Math.* 7, 269–276. (7.1)

Buys, J. D., see Haarhoff and Buys (1970).

Cantrell, J. W., 1969, Relation between the memory gradient method and the Fletcher–Powell method, *J. Optzn. Thry. and Apps.* 4, 67–71. (7.1)

Cantrell, J. W., see Miele and Cantrell (1969, 1970).

Carroll, C. W., 1961, The created response surface technique for optimizing nonlinear restrained systems, *Operations Res.* 9, 169–184. (7.1)

Cauchy, A., 1840, Sur les fonctions interpolaires, *C. R. Acad. Sci. Paris* 11, 775 (or see *Oeuvres complètes*, Gauthier-Villars, Paris, 1897, Vol. 5, 409–424). (2.2)

Cauchy, A., 1847, Méthode générale pour la résolution des systèmes d'équations simultanées, *C. R. Acad. Sci. Paris* 25, 536–538 (or see *Oeuvres complètes*, Gauthier-Villars, Paris, 1897, Vol. 10, 399–402). (7.1)

Chazan, D., and Miranker, W. L., 1970, A non-gradient and parallel algorithm for unconstrained minimization, *SIAM J. Control* 8, 207–217. (7.1, 7.3)

Chernousko, F. L., 1970, On optimal algorithms for search, in Dold and Eckmann (1970a). (4.1)

Clark, N. A., Cody, W. E., Hillstrom, K. E., and Thieleker, E. A., 1967, *Performance statistics of the FORTRAN IV (H) library for the IBM System/360*, Argonne Nat. Lab. Report ANL–7321. (6.3)

Cody, W. J., see Clark, Cody, Hillstrom, and Thieleker (1967).

Collatz, L., 1964, *Functional analysis and numerical mathematics*, Springer-Verlag, Berlin (translation by H. Oser, Academic Press, New York, 1966). (3.1)

Colville, A. R., 1968, *A comparative study of nonlinear programming codes*, IBM New York Scientific Center Report 320-2949. (7.1, 7.7, 7.9)

Conte, S. D., see Brown and Conte (1967).

Cooper, L., see Krolak and Cooper (1963).

Cox, M. G., 1970, A bracketing technique for computing a zero of a function, *Comp. J.* 13, 101–102. (4.2, 4.5)

Cragg, E. E., and Levy, A. V., 1969, Study of a supermemory gradient method for the minimization of functions, *J. Optzn. Thry. and Apps.* 4, 191. (7.1)

Crowder, H., and Wolfe, P., 1971, *Linear convergence of the conjugate gradient method*, Report RC3330, IBM T. J. Watson Research Lab., Yorktown Heights, New York, to appear in *IBM Jour. Res. and Dev.* (7.1, 7.4)

Curry, H., 1944, The method of steepest descent for nonlinear minimization problems, *Quart. Appl. Math.* 2, 258–261. (7.1)

Daniel, J. W., 1967a, The conjugate gradient method for linear and nonlinear operator equations, *SIAM J. Numer. Anal.* 4, 10–26. (7.1)

Daniel, J. W., 1967b, Convergence of the conjugate gradient method with computationally convenient modifications, *Numer. Math.* 10, 125–131. (7.1)

Daniel, J. W., 1970, A correction concerning the convergence rate for the conjugate gradient method, *SIAM J. Numer. Anal.* 7, 277–280. (7.1)

Davidon, W. C., 1959, *Variable metric method for minimization*, Argonne Nat. Lab. Report ANL–5990. (5.7, 7.1)

Davidon, W. C., 1968, Variance algorithm for minimization, *Comp. J.* 10, 406–410. (7.1, 7.7)

Davidon, W. C., 1969, Variance algorithms for minimization, in Fletcher (1969a). (7.1, 7.7)

Davies, D., see Box, Davies, and Swann (1969), Matthews and Davies (1971), Swann (1964).

Davis, P. J., 1965, *Interpolation and approximation*, 2nd ed., Blaisdell, New York and London. (6.2)

Dejon, B., and Henrici, P. (eds.), 1969, *Constructive aspects of the fundamental theorem of algebra*, Interscience, New York.

Dekker, T. J. (ed.), 1963, *The series AP200 of procedures in ALGOL 60*, The Mathematical Centre, Amsterdam.

Dekker, T. J., 1969, Finding a zero by means of successive linear interpolation, in Dejon and Henrici (1969). (1.2, 4.1, 4.2, 4.3, 4.4)

Dekker, T. J., see van Wijngaarden, Zonneveld, and Dijkstra (1963).

Dennis, J. E., 1968, On Newton-like methods, *Numer. Math.* 11, 324–330. (7.1)

Dennis, J. E., 1969a, *On the local convergence of Broyden's method for nonlinear systems of equations*, Tech. Report 69–46, Dept. of Computer Science, Cornell Univ. (7.1)

Dennis, J. E., 1969b, *On the convergence of Broyden's method for nonlinear systems of equations*, Report 69–48, Dept. of Computer Science, Cornell Univ., to appear in *Math. Comp.* (7.1)

Dennis, J. E., see Brown and Dennis (1968, 1971a, b).

Dijkstra, E. W., see van Wijngaarden, Zonneveld, and Dijkstra (1963).

Dixon, L. C. W., 1971a, *Variable metric algorithms: necessary and sufficient conditions for identical behaviour on non-quadratic functions*, Report 26, Numerical Optimisation Centre, The Hatfield Polytechnic. (7.1)

Dixon, L. C. W., 1971b, All the quasi-Newton family generate identical points, to appear. (7.1)

Dold, A., and Eckmann, B. (eds.), 1970a, *Colloquium on methods of optimization* (*Novisibirsk, June 1968*), Springer-Verlag, Berlin. (7.1)

Dold, A., and Eckmann, B., 1970b, see *Symposium on optimization* (*Nice, June 1969*), Springer-Verlag, Berlin. (7.1)

Dreyfus, S. E., see Bellman and Dreyfus (1962).

Dyer, P., see Hanson and Dyer (1971).

Eckmann, B., see Dold and Eckmann (1970a, b).

Ehrlich, L. W., 1970, Eigenvalues of symmetric five-diagonal matrices, unpublished. (4.4)

Evans, J. P., and Gould, F. J., 1970, Stability in nonlinear programming, *Oper. Res.* 18, 107–118. (7.1)

Fiacco, A. V., 1961, Comments on the paper of C. W. Carroll, *Oper. Res.* 9, 184. (7.1)

Fiacco, A. V., 1969, A general regularized sequential unconstrained minimization technique, *SIAM J. Appl. Math.* 17, 1239–1245. (7.1)

Fiacco, A. V., and Jones, A. P., 1969, Generalized penalty methods in topological spaces, *SIAM J. Appl. Math.* 17, 996–1000. (7.1)

Fiacco, A. V., and McCormick, G. P., 1968, *Nonlinear programming: sequential unconstrained minimization techniques*, Wiley, New York. (7.1)

Flanagan, P. D., Vitale, P. A., and Mendelsohn, J., 1969, A numerical investigation of several one-dimensional search procedures in nonlinear regression problems, *Technometrics* 11, 265–284. (5.4)

Fletcher, R., 1965, Function minimization without evaluating derivatives—a review, *Comp. J.* 8, 33–41. (1.2, 7.1, 7.3, 7.5, 7.7, 7.9)

Fletcher, R., 1966, Certification of Algorithm 251, *Comm. ACM* 9, 686. (7.1)

Fletcher, R., 1968a, Generalized inverse methods for the best least squares solution of systems of non-linear equations, *Comp. J.* 10, 392–399. (7.1)

Fletcher, R., 1968b, *Programming under linear equality and inequality constraints*, ICI Management Services Report MSDH/68/19. (7.1)

Fletcher, R. (ed.), 1969a, *Optimization*, Academic Press, New York. (7.1)

Fletcher, R., 1969b, *A class of methods for nonlinear programming with termination and convergence properties*, Report TP 386, AERE, Harwell, England. (7.1)

Fletcher, R., 1969c, A review of methods for unconstrained optimization, in Fletcher (1969a). (7.1, 7.5)

Fletcher, R., 1969d, A technique for orthogonalization, *J. Inst. Maths. Apps.* 5, 162–166. (7.5)

Fletcher, R., 1970, A new approach to variable metric algorithms, *Comp. J.* 13, 317–322. (7.1)

Fletcher, R., and Powell, M. J. D., 1963, A rapidly convergent descent method for minimization, *Comp. J.* 6, 163–168. (7.1, 7.7, 7.9)

Fletcher, R., and Reeves, C. M., 1964, Function minimization by conjugate gradients, *Comp. J.* 7, 149–154. (5.4, 7.1, 7.4)

Forsythe, G. E., 1968, On the asymptotic directions of the s-dimensional optimum gradient method, *Numer. Math.* 11, 57–76. (7.1)

Forsythe, G. E., 1969, Remarks on the paper by Dekker, in Dejon and Henrici (1969). (4.1)

Forsythe, G. E., and Moler, C. B., 1967, *Computer solution of linear algebraic systems*, Prentice-Hall, Englewood Cliffs, New Jersey. (7.2)

Fox, L., Henrici, P., and Moler, C. B., 1967, Approximations and bounds for eigenvalues of elliptic operators, *SIAM J. Numer. Anal.* 4, 89–102. (6.1)

Francis, J., 1962, The QR transformation: a unitary analogue to the LR transformation, *Comp. J.* 4, 265–271. (7.4)

Freudenstein, F., and Roth, B., 1963, Numerical solution of systems of nonlinear equations, *J. ACM* 10, 550–556. (7.7)

Gauss, K. F., 1809, Theoria motus corporum caelestium, *Werke*, Vol. 7, Book 2, Sec. 3. (7.1)

Gill, P. E., and Murray, W., 1970, *A numerically stable form of the simplex algorithm*, Tech. Report Maths. 87, NPL, Teddington, England. (7.1)

Goldfarb, D., 1969, Extensions of Davidon's variable metric method to maximization under linear inequality and equality constraints, *SIAM J. Appl. Math.* 17, 739–764. (7.1)

Goldfarb, D., 1970, A family of variable-metric methods derived by variational means, *Math. Comp.* 24, 23–26. (7.1)

Goldfarb, D., and Lapidus, L., 1968, A conjugate gradient method for nonlinear programming problems with linear constraints, *Indust. Eng. Chem. Fundamentals* 7, 142–151. (7.1)

Goldfeld, S. M., Quandt, R. E., and Trotter, H. F., 1968, *Maximization by improved quadratic hill-climbing and other methods*, Econometrics Research Program Res. Mem. 95, Princeton Univ. (7.1)

Goldstein, A. A., 1962, Cauchy's method of minimization, *Numer. Math.* 4, 146–150. (7.1)

Goldstein, A. A., 1965, On steepest descent, *SIAM J. Control* 3, 147–151. (7.1)

Goldstein, A. A., and Price, J. F., 1967, An effective algorithm for minimization, *Numer. Math.* 10, 184–189. (7.1, 7.7)

Goldstein, A. A., and Price, J. F., 1971, On descent from local minima, *Math. Comp.* 25, 569–574. (6.1)

Golub, G. H., 1965, Numerical methods for solving linear least squares problems, *Numer. Math.* 7, 206–216. (7.1)

Golub, G. H., see Businger and Golub (1965), Bartels and Golub (1969), Bartels, Golub, and Saunders (1970).

Golub, G. H., and Kahan, W., 1965, Calculating the singular values and pseudo-inverse of a matrix, *SIAM J. Numer. Anal.* 2, 205–224. (7.4)

Golub, G. H., and Reinsch, C., 1970, Singular value decomposition and least squares solutions, *Numer. Math.* 14, 403–420. (7.1, 7.4, 7.9)

Golub, G. H., and Saunders, M., 1969, *Linear least squares and quadratic programming*, Report CS 134, Computer Sci. Dept., Stanford Univ. (7.1)

Golub, G. H., and Smith, L. B., 1967, *Chebyshev approximation of continuous functions by a Chebyshev system of functions*, Report CS 72, Computer Sci. Dept., Stanford Univ. (5.4)

Golub, G. H., and Wilkinson, J. H., 1966, Note on the iterative refinement of least squares solutions, *Numer. Math.* 9, 139–148. (7.1)

Gould, F. J., see Evans and Gould (1970).

Graham, S., see Bauer, Becker, and Graham (1968).

Greenstadt, J. L., 1967, On the relative efficiencies of gradient methods, *Math. Comp.* 21, 360–367. (1.2, 7.1)

Greenstadt, J. L., 1970, Variations on variable metric methods, *Math. Comp.* 24, 1–22 (appendix by Y. Bard). (7.1)

Gregory, R. T., and Karney, D. L., 1969, *A collection of matrices for testing computational algorithms*, Interscience, New York. (7.7, 7.9)

Gross, O., and Johnson, S. M., 1959, Sequential minimax search for a zero of a convex function, *MTAC* (now *Math. Comp.*) 13, 44–51. (1.2, 4.1)

Haarhoff, P. C., and Buys, J. D., 1970, A new method for the optimization of a nonlinear function subject to nonlinear constraints, *Comp. J.* 13, 178–184. (7.1)

Hadley, G., 1964, *Nonlinear and dynamic programming*, Addison-Wesley, Reading, Massachusetts. (7.1)

Hanson, R. J., 1970, *Computing quadratic programming problems: linear inequality and equality constraints*, Tech. Memo. 240, JPL, Pasadena. (7.1)

Hanson, R. J., and Dyer, P., 1971, A computational algorithm for sequential estimation, *Comp. J.* 14, 285–290. (7.1)

Hartley, H. O., 1961, The modified Gauss–Newton method for fitting of nonlinear regression functions by least squares, *Technometrics* 3, 269–280. (7.1)

Henrici, P., see Dejon and Henrici (1969), Fox, Henrici, and Moler (1967).

Hext, G. R., see Spendley, Hext, and Himsworth (1962).

Hill, I. D., see Pike, Hill, and James (1967).

Hillstrom, K. E., see Clark, Cody, Hillstrom, and Thieleker (1967).

Himsworth, F. R., see Spendley, Hext, and Himsworth (1962).

Hoare, C., see Wirth and Hoare (1966).

Hooke, R., and Jeeves, T. A., 1961, Direct search solution of numerical and statistical problems, *J. ACM* 8, 212–229. (7.1)

Householder, A. S., 1964, *The theory of matrices in numerical analysis*, Blaisdell, New York. (7.4)

Householder, A. S., 1970, *The numerical treatment of a single nonlinear equation*, McGraw-Hill, New York. (3.1)

Huang, H. Y., 1970, Unified approach to quadratically convergent algorithms for function minimization, *J. Optzn. Thry. and Apps.* 5, 405–423. (7.1)

Isaacson, E., and Keller, H. B., 1966, *Analysis of numerical methods*, Wiley, New York. (2.2, 2.4)

Jacoby, S. L. S., Kowalik, J. S., and Pizzo, J. T., 1971, *Iterative methods for nonlinear optimization problems*, Prentice-Hall, Englewood Cliffs, New Jersey, to appear. (5.4, 7.1)

James, F. D., see Pike, Hill, and James (1967).

Jarratt, P., 1967, An iterative method for locating turning points, *Comp. J.* 10, 82–84. (1.2, 3.1, 3.2, 3.6, 3.7, 3.8, 3.9, 5.1)

Jarratt, P., 1968, A numerical method for determining points of inflexion, *BIT* 8, 31–35. (1.2, 3.1, 3.2, 3.6, 3.9)

Jeeves, T. A., see Hooke and Jeeves (1961).

Jenkins, M. A., 1969, *Three-stage variable-shift iterations for the solution of polynomial equations with a posteriori bounds for the zeros*, Report CS 138, Computer Sci. Dept., Stanford Univ. (3.5)

Johnsen, S. E. J., see Allran and Johnsen (1970).

Johnson, I. L., and Myers, G. E., 1967, *One-dimensional minimization using search by golden section and cubic fit methods*, Report N68–18823 (NASA), Manned Spacecraft Center, Houston. (5.7)

Johnson, S. M., 1955, *Best exploration for maximum is Fibonaccian*, RAND Corp. Report RM–1590. (5.3)

Johnson, S. M., see Gross and Johnson (1959), Bellman (1957), Bellman and Dreyfus (1962).

Jones, A. P., 1970, Spiral—a new algorithm for non-linear parameter estimation using least squares, *Comp. J.* 13, 301–308. (7.1)

Jones, A. P., see Fiacco and Jones (1969).

Kahan, W., see Golub and Kahan (1965).

Kantorovich, L. V., and Akilov, G. P., 1959, *Functional analysis in normed spaces*, Moscow (translation by D. Brown, edited by A. Robertson, MacMillan, New York, 1964). (3.1)

Kaplan, J. L., see Mitchell and Kaplan (1968).

Karney, D. L., see Gregory and Karney (1969).

Karp, R. M., and Miranker, W. L., 1968, Parallel minimax search for a maximum, *J. Comb. Thry.* 4, 19–35. (5.7)

Kaupe, A. F., 1964, On optimal search techniques, *Comm. ACM* 7, 38. (6.7)

Keller, H. B., see Isaacson and Keller (1966).

Kempthorne, O., see Shah, Buehler, and Kempthorne (1964).

Kettler, P. C., see Shanno and Kettler (1969).

Kiefer, J., 1953, Sequential minimax search for a maximum, *Proc. Amer. Math. Soc.* 4, 503–506. (1.2)

Kiefer, J., 1957, Optimal sequential search and approximation methods under minimum regularity assumptions, *SIAM J. Appl. Math.* 5, 105–136. (6.7)

Knuth, D. E., 1969, *The art of computer programming*, Vol. 2, Addison-Wesley, Reading, Massachusetts. (7.9)

Kogbetliantz, E. G., 1955, Solution of linear equations by diagonalization of coefficients matrix, *Quart. Appl. Math.* 13, 123–132. (7.4)

Kowalik, J. S., and Osborne, M. R., 1968, *Methods for unconstrained optimization problems*, Elsevier, New York. (1.2, 2.6, 3.7, 5.3, 5.4, 7.1, 7.7, 7.9)

Kowalik, J. S., Osborne, M. R., and Ryan, D. M., 1969, A new method for constrained optimization problems, *Oper. Res.* 17, 973. (7.1)

Kowalik, J. S., see Jacoby, Kowalik, and Pizzo (1971).

Krolak, P. D., 1968, Further extensions of Fibonaccian search to nonlinear programming problems, *SIAM J. Control* 6, 258–265. (5.3)

Krolak, P. D., and Cooper, L., 1963, An extension of Fibonaccian search to several variables, *Comm. ACM* 6, 639. (6.7)

Kublanovskaya, V. N., 1961, On some algorithms for the solution of the complete eigenvalue problem, *Zh. Vych. Mat.* 1, 555–570. (7.4)

Künzi, H. P., and Oettli, W., 1970, *Nichtlineare Optimierung: Neuere Verfahren Bibliographie*, Springer-Verlag, Berlin.

Künzi, H. P., Tzschach, H. G., and Zehnder, C. A., 1968, *Numerical methods of mathematical optimization*, Academic Press, New York. (7.1)

Lancaster, P., 1966, Error analysis for the Newton–Raphson method, *Numer. Math.* 9, 55–68. (5.2)

Lapidus, L., see Goldfarb and Lapidus (1968).

Lavi, A., and Vogl T. P. (eds.), 1966, *Recent advances in optimization techniques*, Wiley, New York. (7.1, 8)

Lawson, C. L., 1968, *Bibliography of recent publications in approximation theory with emphasis on computer applications*, Tech. Mem. 201, JPL, Pasadena.

Leon, A., 1966, A comparison of eight known optimizing procedures, in Lavi and Vogl (1966). (7.7, 7.9)

Levenberg, K. A., 1944, A method for the solution of certain non-linear problems in least squares, *Quart. Appl. Math.* 2, 164–168. (7.1)

Levy, A. V., see Cragg and Levy (1969).

Lill, S. A., 1970, A modified Davidon method for finding the minimum of a function using difference approximations for derivatives, Algorithm 46, *Comp. J.* 13, 111–113. (7.1)

Lootsma, F. A., 1968, *Constrained optimization via penalty functions*, Philips Res. Report 23, 408. (7.1)

Lootsma, F. A., 1970, *Boundary properties of penalty functions for constrained minimization*, thesis, Eindhoven, Holland. (7.1)

Luenberger, D. G., 1969a, *Optimization by vector space methods*, Wiley, New York. (7.1)

Luenberger, D. G., 1969b, Hyperbolic pairs in the method of conjugate gradients, *SIAM J. Appl. Math.* 17, 1263–1267. (7.1)

Luenberger, D. G., 1970, The conjugate residual method for constrained minimization problems, *SIAM J. Numer. Anal.* 7, 390–398. (7.1)

Magee, E. J., 1960, *An empirical investigation of procedures for locating the maximum peak of a multiple-peak regression function*, Lincoln Lab. Report 22G-0046. (1.2)

Mangasarian, O. L., 1969, *Nonlinear programming*, McGraw-Hill, New York. (7.1)

Marquardt, D. W., 1963, An algorithm for least squares estimation of nonlinear parameters, *J. SIAM* 11, 431–441. (7.1)

Martin, R. S., see Bowdler, Martin, Reinsch, and Wilkinson (1968).

Martin, R. S., Reinsch, C., and Wilkinson, J. H., 1968, Householder's tridiagonalization of a symmetric matrix, *Numer. Math.* 11, 181–195. (7.4)

Matthews, A., and Davies, D., 1971, A comparison of modified Newton methods for unconstrained optimization, *Comp. J.* 14, 293–294. (7.1)

McCormick, G. P., 1969, *The rate of convergence of the reset Davidon variable metric method*, MRC Report 1012, Univ. of Wisconsin. (1.2, 7.1, 7.8)

McCormick, G. P., see Fiacco and McCormick (1968).

McCormick, G. P., and Pearson, J. D., 1969, Variable metric methods and unconstrained optimization, in Fletcher (1969a). (1.2, 7.1, 7.7, 7.9)

Mead, R., see Nelder and Mead (1965).

Meinardus, G., 1967, *Approximation of functions: theory and numerical methods*, Springer-Verlag, Berlin. (3.7)

Mendelsohn, J., see Flanagan, Vitale, and Mendelsohn (1969).

Miele, A., and Cantrell, J. W., 1969, Study of a memory gradient method for the minimization of functions, *J. Optzn. Thry. and Apps.* 3, 459–470. (7.1)

Milne, W. E., 1949, *Numerical calculus*, Princeton Univ. Press, Princeton, New Jersey. (2.2)

Milne-Thomson, L. M., 1933, *The calculus of finite differences*, Macmillan, London. (2.2)

Miranker, W. L., 1969, Parallel methods for approximating the root of a function, *IBM Jour. Res. and Dev.* 13, 297–301. (4.5, 5.7)

Miranker, W. L., see Chazan and Miranker (1970), Karp and Miranker (1968).

Mitchell, R. A., and Kaplan, J. L., 1968, Nonlinear constrained optimization by a non-random complex method, *J. Res. NBS* (*Engr. and Instr.*) 72C, 249. (7.1)

Moler, C. B., see Forsythe and Moler (1967), Fox, Henrici, and Moler (1967).

Murray, W., 1969, Ill-conditioning in barrier and penalty functions arising in constrained nonlinear programming, in *Proceedings of the sixth international symposium on mathematical programming*, Princeton, New Jersey, 1967. (7.1)

Murray, W., see Gill and Murray (1970).

Murtagh, B. A., and Sargent, R. W. H., 1970, Computational experience with quadratically convergent minimization methods, *Comp. J.* 13, 185–194. (7.1)

Myers, G. E., 1968, Properties of the conjugate gradient and Davidon methods, *J. Optzn. Thry. and Apps.* 2, 209–219. (7.1)

Myers, G. E., see Johnson and Myers (1967).

Naur, P. (ed.), 1963, Revised report on the algorithmic language ALGOL 60, *Comm. ACM* 6, 1–17. (1.1)

Nelder, J. A., and Mead, R., 1965, A simplex method for function minimization, *Comp. J.* 7, 308–313. (7.1, 7.4)

Newman, D. J., 1965, Location of the maximum on unimodal surfaces, *J. ACM* 12, 395–398. (1.2, 5.3, 6.7)

Oettli, W., see Künzi and Oettli (1970).

Ortega, J. M., 1968, The Newton–Kantorovich theorem, *Amer. Math. Monthly* 75, 658–660. (3.1)

Ortega, J. M., and Rheinboldt, W. C., 1970, *Iterative solution of nonlinear equations in several variables*, Academic Press, New York. (3.1, 3.2, 3.6, 7.1)

Osborne, M. R., 1969, A note on Powell's method for calculating orthogonal vectors, *Austral. Comp. J.* 1, 216. (7.5)

Osborne, M. R., see Kowalik and Osborne (1968), Kowalik, Osborne, and Ryan (1969).

Osborne, M. R., and Ryan, D. M., 1970, *An algorithm for nonlinear programming*, Report 35, Computer Centre, Australian National Univ., Canberra. (7.1)

Osborne, M. R., and Ryan, D. M., 1971, On penalty function methods for nonlinear programming problems, *J. Math. Anal. Apps.*, to appear. (7.1)

Ostrowski, A. M., 1966, *Solution of equations and systems of equations*, Academic Press, New York (2nd edition). (1.2, 3.1, 3.2, 3.6, 3.7, 4.2, 5.1, 7.1))

Ostrowski, A. M., 1967a, Contributions to the theory of the method of steepest descent, *Arch. Rational Mech. Anal.* 26, 257–280. (7.1)

Ostrowski, A. M., 1967b, The round-off stability of iterations, *Z. Angew. Math. Mech.* 47, 77–82. (5.2)

Overholt, K. J., 1965, An instability in the Fibonacci and the golden section search methods, *BIT* 5, 284. (5.3)

Overholt, K. J., 1967, Note on Algorithm 2, Algorithm 16 and Algorithm 17, *Comp. J.* 9, 414. (5.3)

Palmer, J. R., 1969, An improved procedure for orthogonalising the search vectors in Rosenbrock's and Swann's direct search optimization methods, *Comp. J.* 12, 69. (7.5)

Parlett, B. N., 1971, Analysis of algorithms for reflections in bisectors, *SIAM Review* 13, 197–208. (7.4)

Pearson, J. D., 1969, Variable metric methods of minimization, *Comp. J.* 12, 171–178. (7.1)

Pearson, J. D., see McCormick and Pearson (1969).

Peckham, G., 1970, A new method for minimizing a sum of squares without calculating gradients, *Comp. J.* 13, 418–420. (7.1)

Peters, G., and Wilkinson, J. H., 1969, Eigenvalues of $Ax = \lambda Bx$ with band symmetric A and B, *Comp. J.* 12, 398–404. (1.2, 4.1, 4.2)

Pierre, D. A., 1969, *Optimization theory with applications*, Wiley, New York. (5.4)

Pietrzykowski, T., 1969, An exact potential method for constrained maxima, *SIAM J. Numer. Anal.* 6, 229. (7.1)

Pike, M. C., Hill, I. D., and James, F. D., 1967, Note on Algorithm 2, Fibonacci Search and on Algorithm 7, MINIX, *Comp. J.* 9, 416. (5.2)

Pike, M. C., and Pixner, J., 1967, Algorithm 2, Fibonacci Search, *Comp. Bulletin* 8, 147. (5.3)

Pike, M. C., see Bell and Pike (1966).

Pixner, J., see Pike and Pixner (1967).

Pizzo, J. T., see Jacoby, Kowalik, and Pizzo (1971).

Powell, M. J. D., 1962, An iterative method for finding stationary values of a function of several variables, *Comp. J.* 5, 147–151. (7.7, 7.9)

Powell, M. J. D., 1964, An efficient method for finding the minimum of a function of several variables without calculating derivatives, *Comp. J.* 7, 155–162. (1.1, 1.2, 5.4, 7.1, 7.3, 7.5, 7.6, 7.7, 7.8, 7.9)

Powell, M. J. D., 1965, A method of minimizing a sum of squares of non-linear functions without calculating derivatives, *Comp. J.* 7, 303–307. (7.1, 7.7)

Powell, M. J. D., 1966, Minimization of functions of several variables, in Walsh (1966). (7.1)

Powell, M. J. D., 1968a, On the calculation of orthogonal vectors, *Comp. J.* 11, 302–304. (7.5)

Powell, M. J. D., 1968b, *A FORTRAN subroutine for solving systems of non-linear equations*, Report R–5947, AERE, Harwell, England. (7.1)

Powell, M. J. D., 1969a, *A hybrid method for nonlinear equations*, Report TP 364, AERE, Harwell, England. (7.1)

Powell, M. J. D., 1969b, *On the convergence of the variable metric algorithm*, Report TP 382, AERE, Harwell, England. (7.1)

Powell, M. J. D., 1969c, A theorem on rank one modifications to a matrix and its inverse, *Comp. J.* 12, 288–290. (7.1)

Powell, M. J. D., 1970a, A survey of numerical methods for unconstrained optimization, *SIAM Review* 12, 79–97. (7.1)

Powell, M. J. D., 1970b, *A new algorithm for unconstrained optimization*, Report TP 393, AERE, Harwell, England. (7.1)

Powell, M. J. D., 1970c, Rank one methods for unconstrained optimization, in Abadie (1970). (7.1)

Powell, M. J. D., 1970d, *A FORTRAN subroutine for unconstrained minimization, requiring first derivatives of the objective function*, Report R–6469, AERE, Harwell, England. (7.1)

Powell, M. J. D., 1970e, *Recent advances in unconstrained optimization*, Report TP 430, AERE, Harwell, England. (7.1, 7.7)

Powell, M. J. D., see Fletcher and Powell (1963).

Price, J. F., see Goldstein and Price (1967, 1971).

Quandt, R. E., see Goldfeld, Quandt, and Trotter (1968).

Rall, L. B. (ed.), 1965, *Error in digital computation*, Vol. 2, Wiley, New York.

Rall, L. B., 1966, Convergence of the Newton process to multiple solutions, *Numer. Math.* 9, 23–37. (7.1)

Rall, L. B., 1969, *Computational solution of nonlinear operator equations*, Wiley, New York. (7.1)

Ralston, A., 1963, On differentiating error terms, *Amer. Math. Monthly* 70, 187–188. (1.2, 2.1, 2.6)

Ralston, A., 1965, *A first course in numerical analysis*, McGraw-Hill, New York. (1.2, 2.6)

Ralston, A., and Wilf, H. S. (eds.), 1960, *Mathematical methods for digital computers*, Vol. 1, Wiley, New York. (7.1)

Ralston, A., and Wilf, H. S. (eds.), 1967, *Mathematical methods for digital computers*, Vol. 2, Wiley, New York.

Ramsay, J. O., 1970, A family of gradient methods for optimization, *Comp. J.* 13, 413–417. (7.1)

Reeves, C. M., see Fletcher and Reeves (1964).

Reinsch, C., see Golub and Reinsch (1970), Martin, Reinsch, and Wilkinson (1968), Bowdler, Martin, Reinsch, and Wilkinson (1968).

Rhead, D. G., 1971, *Some numerical experiments on Zangwill's method for unconstrained minimization*, Working Paper ICSI 319, Univ. of London. (7.3)

Rheinboldt, W. C., see Ortega and Rheinboldt (1970).

Rice, J. R., 1970, Minimization and techniques in nonlinear approximation, *SIAM Studies in Numer. Anal.* 2, 80–98. (7.1)

Richman, P. L., 1968, *ϵ-calculus*, Report CS 105, Stanford Univ. (1.2, 5.3)

Rivlin, T. J., 1970, Bounds on a polynomial, *J. Res. NBS* 74B, 47–54. (1.2, 6.1)

Robbins, H., 1952, Some aspects of the sequential design of experiments, *Bull. Amer. Math. Soc.* 58, 527–536. (1.2)

Rosen, J. B., 1960, The gradient projection method for nonlinear programming. Part 1: Linear constraints, *J. SIAM* 8, 181. (7.1)

Rosen, J. B., 1961, The gradient projection method for nonlinear programming. Part 2: Nonlinear constraints, *J. SIAM* 9, 514. (7.1)

Rosen, J. B., and Suzuki, S., 1965, Construction of nonlinear programming test problems, *Comm. ACM* 8, 113. (7.1)

Rosenbrock, H. H., 1960, An automatic method for finding the greatest or least value of a function, *Comp. J.* 3, 175–184. (6.8, 7.5, 7.7, 7.9)

Roth, B., see Freudenstein and Roth (1963).

Ryan, D. M., see Osborne and Ryan (1970, 1971), Kowalik, Osborne, and Ryan (1969).

Sargent, R. W. H., see Murtagh and Sargent (1969, 1970).

Satterthwaite, E., see Bauer, Becker, and Graham (1968).

Saunders, M., see Golub and Saunders (1969), Bartels, Golub, and Saunders (1970).

Schröder, E., 1870, Über unendlich viele Algorithmen zur Auflösung der Gleichungen, *Math. Ann.* 2, 317–365. (3.1)

Schubert, L. K., 1970, Modification of a quasi-Newton method for nonlinear equations with a sparse Jacobian, *Math. Comp.* 24, 27–30. (7.1)

Shah, B. V., Buehler, R. J., and Kempthorne, O., 1964, Some algorithms for minimizing a function of several variables, *SIAM J. Appl. Math.* 12, 74–92. (7.1)

Shanno, D. F., 1970a, Parameter selection for modified Newton methods for function minimization, *SIAM J. Numer. Anal.* 7, 366–372. (7.1)

Shanno, D. F., 1970b, An accelerated gradient projection method for linearly constrained nonlinear estimation, *SIAM J. Appl. Math.* 18, 322–334. (7.1)

Shanno, D. F., and Kettler, P. C., 1969, *Optimal conditioning of quasi-Newton methods*, Center for Math. Studies in Business and Economics Report 6937, Univ. of Chicago. (7.1)

Smith, C. S., 1962, *The automatic computation of maximum likelihood estimates*, NCB Sci. Dept. Report SC 846/MR/40. (7.1, 7.3)

Smith, L. B., see Golub and Smith (1967).

Sobel, I., 1970, *Camera models and machine perception*, Stanford Artificial Intelligence Report AIM–121. (7.7)

Sorensen, H. W., 1969, Comparison of some conjugate direction procedures for function minimization, *J. Franklin Institute* 288, 421. (7.1)

Spang, H. A., 1962, A review of minimization techniques for nonlinear functions, *SIAM Review* 4, 343–365. (7.1)

Späth, H., 1967, The damped Taylor series method for minimizing a sum of squares and for solving systems of nonlinear equations, *Comm. ACM* 10, 726–728. (7.1)

Spendley, W., Hext, G. R., and Himsworth, F. R., 1962, Sequential application of simplex designs in optimization and evolutionary operation, *Technometrics* 4, 441. (7.1)

Sproull, R., see Swinehart and Sproull (1970).

Stewart, G. W., 1967, A modification of Davidon's minimization method to accept difference approximations of derivatives, *J. ACM* 14, 72–83. (1.1, 1.2, 7.1, 7.7, 7.8)

Stiefel, E. L., see Hestenes and Stiefel (1952).

Stoer, J., 1971, On the numerical solution of constrained least squares problems, *SIAM J. Numer. Anal.* 8, 382–411. (7.1)

Sugie, N., 1964, An extension of Fibonaccian searching to multidimensional cases, *IEEE Trans. Control* AC–9, 105. (6.7)

Suzuki, S., see Rosen and Suzuki (1965).

Swann, W. H., 1964, *Report on the development of a new direct search method of optimization*, ICI Ltd. Cent. Inst. Lab. Research Note 64/3. (1.2, 7.1, 7.5)

Swann, W. H., see Box, Davies, and Swann (1969).

Swinehart, D., and Sproull, R., 1970, *SAIL*, Stanford Artificial Intelligence Project Operating Note 57.1. (7.7)

Takahashi, I., 1965, A note on the conjugate gradient method, *Information Processing in Japan* 5, 45–49. (7.1)

Thieleker, E. A., see Clark, Cody, Hillstrom, and Thieleker (1967).

Tornheim, L., 1964, Convergence of multipoint iterative methods, *J. ACM* 11, 210–220. (3.2)

Traub, J. F., 1964, *Iterative methods for the solution of equations*, Prentice-Hall, Englewood Cliffs, New Jersey. (2.2, 3.1, 3.2, 4.5)

Traub, J. F., 1967, The solution of transcendental equations, in Ralston and Wilf (1967). (3.1, 3.2)

Trotter, H. F., see Goldfeld, Quandt, and Trotter (1968).

Tzschach, H. G., see Künzi, Tzschach, and Zehnder (1968).

Vitale, P. A., see Flanagan, Vitale, and Mendelsohn (1969).

Vogl, T. P., see Lavi and Vogl (1966).

Voigt, R. G., 1971, Orders of convergence for iterative procedures, *SIAM J. Numer. Anal.* 8, 222–243. (3.2, 7.1)

Wall, D., 1956, The order of an iteration formula, *Math. Comp.* 10, 167–168. (3.2)

Walsh, J. (ed.), 1966, *Numerical analysis: an introduction*, Academic Press, New York.

Wells, M., 1965, Algorithm 251, Function minimization, *Comm. ACM* 8, 169–170. (7.1)

van Wijngaarden, A., Zonneveld, J. A., and Dijkstra, E. W., 1963, Programs AP200 and AP230 de serie AP200, in Dekker (1963). (1.2, 4.1)

Wilde, D. J., 1964, *Optimum seeking methods*, Prentice-Hall, Englewood Cliffs, New Jersey. (1.2, 4.5, 5.3, 5.7, 7.1, 7.5)

Wilde, D. J., and Beightler, C. S., 1967, *Foundations of optimization*, Prentice-Hall, Englewood Cliffs, New Jersey. (7.1)

Wilde, D. J., see Avriel and Wilde (1966).

Wilf, H. S., see Ralston and Wilf (1960, 1967).

Wilkinson, J. H., 1963, *Rounding errors in algebraic processes*, HMSO, London. (4.2, 6.3, 7.2)

Wilkinson, J. H., 1965a, *The algebraic eigenvalue problem*, Oxford Univ. Press, Oxford. (7.2, 7.4)

Wilkinson, J. H., 1965b, Error analysis of transformations based on the use of matrices of the form I-2ww^H, in Rall (1965). (7.4)

Wilkinson, J. H., 1967, *Two algorithms based on successive linear interpolation*, Report CS 60, Computer Sci. Dept., Stanford Univ. (1.2, 4.1, 4.2)

Wilkinson, J. H., 1968, Global convergence of the QR algorithm, in *Proceedings of IFIPS Congress (Edinburgh, 1968)*. (7.4)

Wilkinson, J. H., see Peters and Wilkinson (1969), Golub and Wilkinson (1966), Martin, Reinsch, and Wilkinson (1968), Bowdler, Martin, Reinsch, and Wilkinson (1968).

Winfield, D. H., 1967, *Function minimization without derivatives by a sequence of quadratic programming problems*, Report 537, Engineering & Applied Physics Division, Harvard Univ. (7.1)

Winograd, S., and Wolfe, P., 1971, *Optimal iterative processes*, Report RC3511, IBM T. J. Watson Research Lab., Yorktown Heights, New York. (4.5)

Wirth, N., and Hoare, C., 1966, A contribution to the development of ALGOL, *Comm. ACM* 9, 413–431. (1.1, 4.4, 5.6, 6.6, 7.9)

Witzgall, C., 1969, *Fibonacci search with arbitrary first evaluation*, Report D1–82–0916, Boeing Scientific Research Labs., Seattle, Washington. (1.2, 5.3, 5.4)

Wolfe, P., 1959, The secant method for simultaneous non-linear equations, *Comm. ACM* 2, 12–13. (7.1)

Wolfe, P., 1963, Methods of nonlinear programming, in *Recent advances in nonlinear programming*, edited by R. L. Graves and P. Wolfe, McGraw-Hill, New York. (7.1)

Wolfe, P., 1969, Convergence conditions for ascent methods, *SIAM Review* 11, 226–235. (7.1)

Wolfe, P., 1971, Convergence conditions for ascent methods II: some corrections, *SIAM Review* 13, 185–188. (7.1)

Wolfe, P., see Crowder and Wolfe (1971), Winograd and Wolfe (1971).

Zadeh, L. A. (ed.), 1969, *Computing methods in optimization problems*, Vol. 2, Academic Press, New York. (7.1)

Zangwill, W. I., 1967a, Minimizing a function without calculating derivatives, *Comp. J.* 10, 293–296. (7.1, 7.3)

Zangwill, W. I., 1967b, Nonlinear programming via penalty functions, *Mgmt. Sci.* 13, 344–358. (7.1)

Zangwill, W. I., 1969a, *Nonlinear programming: a unified approach*, Prentice-Hall, Englewood Cliffs, New Jersey. (7.1)

Zangwill, W. I., 1969b, Convergence conditions for nonlinear programming algorithms, *Mgmt. Sci.* 16, 1. (7.1)

Zehnder, C. A., see Künzi, Tzschach, and Zehnder (1968).

Zeleznik, F. J., 1968, Quasi-Newton methods for nonlinear equations, *J. ACM* 15, 265–271. (7.1)

Zonneveld, J. A., see van Wijngaarden, Zonneveld, and Dijkstra (1963).

Zoutendijk, G., 1960, *Methods of feasible directions*, Elsevier, Amsterdam and New York. (7.1)

Zoutendijk, G., 1966, Nonlinear programming: a numerical survey, *SIAM J. Control* 4, 194–210. (7.1)

APPENDIX

This Appendix contains FORTRAN translations of the ALGOL 60 procedures *zero*, *localmin*, and *glomin* given in Sections 4.6, 5.8, and 6.10. The FORTRAN subroutines follow the ALGOL procedures as closely as possible, and have been tested with a FORTRAN H compiler on an IBM 360/91 computer.

```fortran
C        A FORTRAN TRANSLATION OF THE ALGOL PROCEDURE ZERO.
C        SEE PROCEDURE ZERO, SECTION 4.6, FOR COMMENTS ETC.
         REAL FUNCTION ZERO (A, B, MACHEP, T, F)
         REAL A,B,MACHEP,T,F,SA,SB,C,D,E,FA,FB,FC,TOL,M,P,Q,R,S
         SA = A
         SB = B
         FA = F(SA)
         FB = F(SB)
   10 C = SA
         FC = FA
         E = SB - SA
         D = E
   20 IF (ABS(FC).GE.ABS(FB)) GO TO 30
         SA = SB
         SB = C
         C = SA
         FA = FB
         FB = FC
         FC = FA
   30 TOL = 2.0*MACHEP*ABS(SB) + T
         M = 0.5*(C - SB)
         IF ((ABS(M).LE.TOL).OR.(FB.EQ.0.0)) GO TO 140
         IF ((ABS(E).GE.TOL).AND.(ABS(FA).GT.ABS(FB))) GO TO 40
         E = M
         D = E
         GO TO 100
   40 S = FB/FA
         IF (SA.NE.C) GO TO 50
         P = 2.0*M*S
         Q = 1.0 - S
         GO TO 60
   50 Q = FA/FC
         R = FB/FC
         P = S*(2.0*M*Q*(Q - R) - (SB - SA)*(R - 1.0))
         Q = (Q - 1.0)*(R - 1.0)*(S - 1.0)
   60 IF (P.LE.0.0) GO TO 70
         Q = -Q
         GO TO 80
   70 P = -P
   80 S = E
         E = D
         IF ((2.0*P.GE.3.0*M*Q-ABS(TOL*Q)).OR.(P.GE.ABS(0.5*S*Q))) GO TO 90
         D = P/Q
         GO TO 100
   90 E = M
         D = E
  100 SA = SB
         FA = FB
         IF (ABS(D).LE.TOL) GO TO 110
         SB = SB + D
         GO TO 130
  110 IF (M.LE.0.0) GO TO 120
         SB = SB + TOL
         GO TO 130
  120 SB = SB - TOL
  130 FB = F(SB)
         IF ((FB.GT.0.0).AND.(FC.GT.0.0)) GO TO 10
         IF ((FB.LE.0.0).AND.(FC.LE.0.0)) GO TO 10
         GO TO 20
  140 ZERO = SB
         RETURN
         END
C        A FORTRAN TRANSLATION OF THE ALGOL PROCEDURE LOCALMIN.
C        SEE PROCEDURE LOCALMIN, SECTION 5.8, FOR COMMENTS ETC.
         REAL FUNCTION LOCALM (A, B, EPS, T, F, X)
         REAL A,B,EPS,T,F,X,SA,SB,D,E,M,P,Q,R,TOL,T2,U,V,W,FU,FV,FW,FX
```

```
      SA = A
      SB = B
      X = SA + 0.381966*(SB - SA)
      W = X
      V = W
      E = 0.0
      FX = F(X)
      FW = FX
      FV = FW
10    M = 0.5*(SA + SB)
      TOL = EPS*ABS(X) + T
      T2 = 2.0*TOL
      IF (ABS(X-M).LE.T2-0.5*(SB-SA)) GO TO 190
      R = 0.0
      Q = R
      P = Q
      IF (ABS(E).LE.TOL) GO TO 40
      R = (X - W)*(FX - FV)
      Q = (X - V)*(FX - FW)
      P = (X - V)*Q - (X - W)*R
      Q = 2.0*(Q - R)
      IF (Q.LE.0.0) GO TO 20
      P = -P
      GO TO 30
20    Q = -Q
30    R = E
      E = D
40    IF (ABS(P).GE.ABS(0.5*Q*R)) GO TO 60
      IF ((P.LE.Q*(SA-X)).OR.(P.GE.Q*(SB-X))) GO TO 60
      D = P/Q
      U = X + D
      IF ((U-SA.GE.T2).AND.(SB-U.GE.T2)) GO TO 90
      IF (X.GE.M) GO TO 50
      D = TOL
      GO TO 90
50    D = -TOL
      GO TO 90
60    IF (X.GE.M) GO TO 70
      E = SB - X
      GO TO 80
70    E = SA - X
80    D = 0.381966*E
90    IF (ABS(D).LT.TOL) GO TO 100
      U = X + D
      GO TO 120
100   IF (D.LE.0.0) GO TO 110
      U = X + TOL
      GO TO 120
110   U = X - TOL
120   FU = F(U)
      IF (FU.GT.FX) GO TO 150
      IF (U.GE.X) GO TO 130
      SB = X
      GO TO 140
130   SA = X
140   V = W
      FV = FW
      W = X
      FW = FX
      X = U
      FX = FU
      GO TO 10
150   IF (U.GE.X) GO TO 160
      SA = U
      GO TO 170
160   SB = U
```

```
  170 IF ((FU.GT.FW).AND.(W.NE.X)) GO TO 180
      V = W
      FV = FW
      W = U
      FW = FU
      GO TO 10
  180 IF ((FU.GT.FV).AND.(V.NE.X).AND.(V.NE.W)) GO TO 10
      V = U
      FV = FU
      GO TO 10
  190 LOCALM = FX
      RETURN
      END
C     A FORTRAN TRANSLATION OF THE ALGOL PROCEDURE GLOMIN.
C     SEE PROCEDURE GLOMIN, SECTION 6.10, FOR COMMENTS ETC.
      REAL FUNCTION GLOMIN (A, B, C, M, MACHEP, E, T, F, X)
      REAL A,B,C,M,MACHEP,E,T,F,X,SC
      REAL A0,A2,A3,D0,D1,D2,H,M2,P,Q,QS,R,S,Y,Y0,Y1,Y2,Y3,YB,Z0,Z1,Z2
      INTEGER K
      A0 = B
      X = A0
      A2 = A
      Y0 = F(B)
      YB = Y0
      Y2 = F(A)
      Y = Y2
      IF (Y0.GE.Y) GO TO 10
      Y = Y0
      GO TO 20
   10 X = A
   20 IF ((M.LE.0.0).OR.(A.GE.B)) GO TO 140
      M2 = 0.5*(1.0 + 16.0*MACHEP)*M
      SC = C
      IF ((SC.LE.A).OR.(SC.GE.B)) SC = 0.5*(A + B)
      Y1 = F(SC)
      K = 3
      D0 = A2 - SC
      H = 0.8181818
      IF (Y1.GE.Y) GO TO 30
      X = SC
      Y = Y1
   30 D1 = A2 - A0
      D2 = SC - A0
      Z2 = B - A2
      Z0 = Y2 - Y1
      Z1 = Y2 - Y0
      R = D1*D1*Z0 - D0*D0*Z1
      P = R
      QS = 2.0*(D0*Z1 - D1*Z0)
      Q = QS
      IF ((K.GT.100000).AND.(Y.LT.Y2)) GO TO 50
   40 IF (Q*(R*(YB-Y2)+Z2*Q*((Y2-Y)+T)).GE.Z2*M2*R*(Z2*Q-R)) GO TO 50
      A3 = A2 + R/Q
      Y3 = F(A3)
      IF (Y3.GE.Y) GO TO 50
      X = A3
      Y = Y3
C     ASSUME THAT 1611*K DOES NOT OVERFLOW.
   50 K = MOD (1611*K, 1048576)
      Q = 1.0
      R = (B - A)*0.00001*FLOAT(K)
      IF (R.LT.Z2) GO TO 40
      R = M2*D0*D1*D2
      S = SQRT ((((Y2 - Y) + T)/M2)
      H = 0.5*(1.0 + H)
      P = H*(P + 2.0*R*S)
      Q = R + 0.5*QS
```

```
      R = -0.5*(DO + (ZO + 2.01*E)/(DO*M2))
      IF ((R.GE.S).AND.(DO.GE.0.0)) GO TO 60
      R = A2 + S
      GO TO 70
   60 R = A2 + R
   70 IF (P*Q.LE.0.0) GO TO 80
      A3 = A2 + P/Q
      GO TO 90
   80 A3 = R
   90 IF (A3.LT.R) A3 = R
      IF (A3.LT.B) GO TO 100
      A3 = B
      Y3 = YB
      GO TO 110
  100 Y3 = F(A3)
  110 IF (Y3.GE.Y) GO TO 120
      X = A3
      Y = Y3
  120 DO = A3 - A2
      IF (A3.LE.R) GO TO 130
      P = 2.0*(Y2 - Y3)/(M*DO)
      IF (ABS(P).GE.(1.0+9.0*MACHEP)*DO) GO TO 130
      IF (0.5*M2*(DO*DO+P*P).LE.(Y2-Y)+(Y3-Y)+2.0*T) GO TO 130
      A3 = 0.5*(A2 + A3)
      H = 0.9*H
      GO TO 90
  130 IF (A3.GE.B) GO TO 140
      AO = SC
      SC = A2
      A2 = A3
      YO = Y1
      Y1 = Y2
      Y2 = Y3
      GU TO 30
  140 GLOMIN = Y
      RETURN
      END
```

INDEX

193